CULTURE AND COSMOS
Vol. 1 No. 1   Spring/Summer 1997   ISSN 1368-6534

Editorial

Astronomy is more than the science of the stars. It is intimately connected to our ideas of our selves, our purpose and place in the universe. Currently it is fueling myths, beliefs and ideologies as much as at any time in its history. We have the neo-paganism based on such sites as Stonehenge and Avebury. There is the belief in UFOs, borne out of modern science and science fiction, which in the last decade has exploded in ever more detailed doctrines of alien visitation and abduction. On a more sober level we have the manipulation of the space race to provide propaganda for the cold war. In the post 1989 world the USA continues to use its space programme to reinforce images of the American dream; space has become the new frontier, and 1997's Mars Rover mission was designed to touch down on 4 July, the anniversary of the declaration of independence. Then there is the social impact of the new science, explored in terms of quantum physics by Danah Zohar and Ian Marshall.[2] If the political impact of Copernicus and Newton is now well established, then we have hardly begun to question what effect Einstein may have had on twentieth century politics.

Then there is the question of belief in astrology which, in the modern world, is perhaps as strong as it ever has been. Its popularity strikes many people as something of a historical problem: how in a modern rational, scientific world could such an irrational, unscientific belief flourish? Given that such words as rational and scientific are highly problematic and open to differing interpretations, astrology's contemporary revival does require study and explanation.

The significance of astrology to the history of ideas, religion and science is no longer in question. Lynn Thorndike's epic *History of Magic and Experimental Science* has made the argument convincingly. A select group of other scholars have examined its role in specific periods and cultures, notably in the ancient near east, the classical world and in medieval and Renaissance Europe. Yet while papers on the subject find their place in specialist academic journals, it is felt that the time has come for a journal which focuses specifically on its past and development, as well as on its social, political and religious functions.

Astrology can be defined as the use of celestial phenomena to interpret and predict events on earth. However, in order to flourish it requires a philosophical context based on a general belief that movements and changes in the heavens are significant for humanity, without the specific rules, technicalities and procedures necessary for astrological interpretation. This may be defined as cultural astronomy: the use of astronomical knowledge, beliefs or theories to inspire, inform or influence social forms and ideologies, or any aspect of human behaviour. Cultural astronomy also includes the modern disciplines of ethnoastronomy and archaeoastronomy.

The problem of definitions was dealt with by Michael Hoskin in his 1996 review of *Astronomies and Cultures*.[1] He posed the question "What astronomy is *not* an astronomy in a culture?" This is a valid point, and one may take a narrow or broad definition of the term cultural astronomy. In part the solution is one of emphasis. *Culture and Cosmos* will emphasise the cultural aspects of astronomy rather than the strict history of mathematical and technical astronomy, areas ably catered for elsewhere.

Culture and Cosmos will cover a broad spectrum of ideas: any and all of the ways in which human beings manipulate, exploit, analyse and interpret the heavens in order to understand, regulate and predict their individual concerns and collective lives. We may not be able to answer all questions, but at least we can ask them.

References

1. Hoskin, Michael, review of *Astronomies and Cultures*, ed. Clive L. N. Ruggles and Nicholas J. Saunders (University of Colorado Press, Niwot, Col., 1993), *Archaeoastronomy*, number 21, supplement to the *Journal for the History of Astronomy*, vol. 27, 1996, p 885-7.

2. Danah Zohar and Ian Marshall, *The Quantum Society*, (Bloomsbury Publishing, London 1993).

An Astronomical Basis for the Myth of the Solar Hero

Robin Heath*

Introduction
Our increasing knowledge of the megalithic culture of the British Isles in the 2nd and 3rd millennia BCE tends to confirm the proposition that megalithic astronomers measured celestial positions with considerable accuracy. The evidence indicates that they understood the 18.6 year nodal period and the moon's nine minute declination wobble.[1] They also had sufficient geometrical ability to re-proportion spacings between lines, divide circles into whole number polygons and divide lines into equal integer spacings.[2] We should therefore ask whether there is evidence of such early astronomy in the numbers which recur in certain myths. The following should be viewed as preliminary arguments.

The Thirty-Three Year Cycle
If our interest is megalithic astronomy then we should search for relevant evidence in the myths of the British Isles. One of the most recurrent numbers in the stories of the Tuatha de Danaan who, according to tradition, inhabited Ireland before 1,500 BCE, is thirty-three.[3] We are told, for example, that the first battle of Mag Tuired was fought by the saviour-hero Lug and thirty-two other leaders. In the same vein, Nemed, another hero, reached Ireland with only one ship, while thirty-three were lost on the way; Cuchulainn slays thirty-three of the Labriads in the Bru battle whilst a late account of the second battle of Mag Tuired names thirty-three leaders of the Fomorii, thirty-two plus their highest king.[4]

This material contains one clear and obvious common theme. Repeatedly, it reinforces an originally oral message which told the knowing listener to look to the number thirty-three as something relevant to a hero, a saviour. In his analysis of the Welsh *White Book of Rhydderch*, N.L.Thomas writes that 'Both three and eleven were equally symbolic, the multiplicand thirty-three particularly so. It has frequently been used to imply supra-human attributes, regal authority and deification.'[5]

---

* Megalithic Tours, Cwm Degwel, St.Dogmael's, Cardigan, SA43 3JF, UK.

We find evidence of the astronomical and mythical significance of the thirty three year cycle in other cultures. Perhaps the best known example is found in the tradition that Christ began his ministry at age thirty and was crucified at age thirty three.[6] The solar tradition in early Christianity is well-recorded, with the widespread identification of Christ with Helios and the fixing of Christmas to coincide with the festival of Sol Invictus, a few days after the winter solstice, and Easter close to the spring equinox.[7] It is partly on the basis of such evidence, together with the argument that epic myths such as those of Gilgamesh and Hercules represent solar cycles, that modern comparative mythology has produced the notion of the solar hero.[8] I am arguing that the numerical evidence in the Celtic tales provides astronomical evidence that they too could be considered solar heroes.

Megalithic Astronomy and the Solar Cycle
Many megalithic standing stones have been shown to relate to extreme Sun and Moon rising and setting azimuths against the local horizon. While the link at some sites is tenuous it is beyond doubt in others, as for example at Stonehenge.[9] The practical solar year is 365 days long. I say *practical* because most reference books claim that there are 365 and a quarter days in the year, a confusion with the Earth's orbital period around the Sun. Every fourth year an extra day slips in to make it 366 days. In four years there are thus 1461 days. It is fairly easy to observe the Sun's behaviour and thereby measure this number.

An equinoctial Sunrise marker, of which many still exist in the British uplands will, each year, deliver the vernal equinox sunrise from a slightly different position on the horizon. The 'quarter day effect' means that the Sun, each year, is displaced about a quarter of a degree from the marker stone. During three years of observation, the Sun appears to be slipping every more away from the alignment until, at the fourth year, two remarkable and very observable things happen simultaneously: the Sun rises once more very close to the marker stone, whence the day count - the tally - for the year is found to be 366 and not 365 days. Observation does not stop there, and a good human eye can detect much more minuscule angular changes than a quarter of a degree from watching sunrises.[10] The truth about solar year measurements carried out at the equinox is that the result is always 365 days unless sustained observations are conducted over many years. In that case the result is 365.25 days, a figure which, under optimum conditions could be reached after four years (see Figure 1).[11] This figure is within eleven minutes of time of that for the tropical year (365.24219 mean solar days) and is almost identical to that for the civil, calendar, year (365.2425 mean solar days).

Figure 1

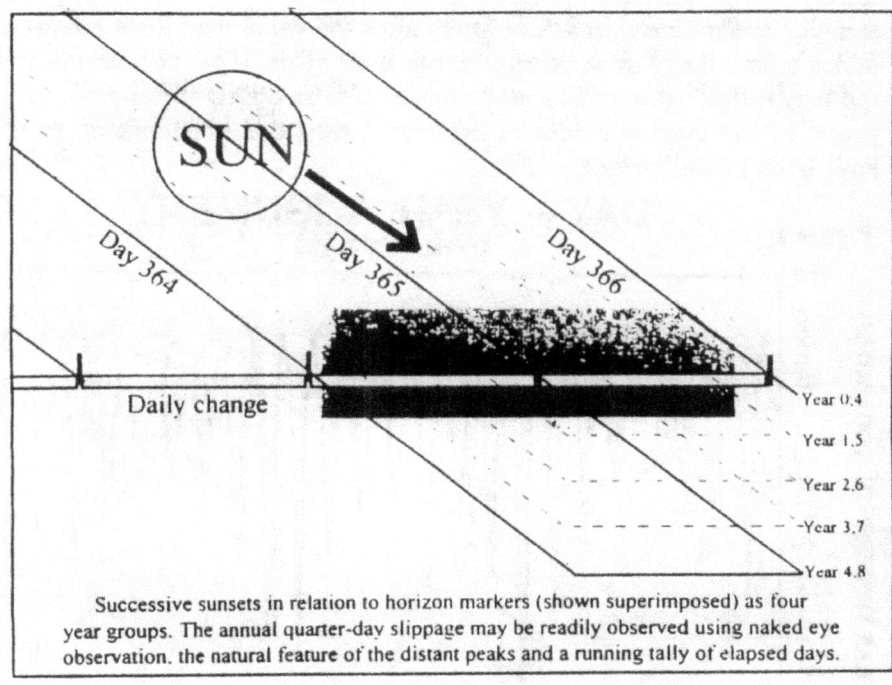

Successive sunsets in relation to horizon markers (shown superimposed) as four year groups. The annual quarter-day slippage may be readily observed using naked eye observation. the natural feature of the distant peaks and a running tally of elapsed days.

For longer time periods something else happens. Every once in a whole number of years the chance arises to measure the year with even greater precision. This can be achieved by observing certain key years when, once again, the Sun rises *precisely* behind the foresight - be this a stone marker or a distant mountain peak - in other words, *a perfect repeat solar cycle*.

What we may assume, courtesy of their enduring architecture, is that the megalithic astronomers could have readily evaluated the length of the solar year to two decimal places. They could accomplish this by marking 1461 equal lengths on a rope - the tally count of days in four years - and then folding it in half twice to obtain 365.25. The 1461 day tally is a given, gleaned from simple observation and tally counting over four years.

After thirty-three years one can observe an <u>exact</u> repeat of the original equinoctial rising behind the marker stone (See Table 1, Figure 2). To a megalithic astronomer, this same phenomenon would have translated as an *exact repeat rising* (or setting) behind a marker, whilst a modern astronomer would note that the Sun's declination will be identical on the same calendar date thirty-three years after the value read from a book of tables today; thirty three years is a true solar cycle. If we can assume that the megalithic astronomers made exact angular observations over many years, as the current evidence suggests,[12] then this phenomenon would have been a familiar one to them.

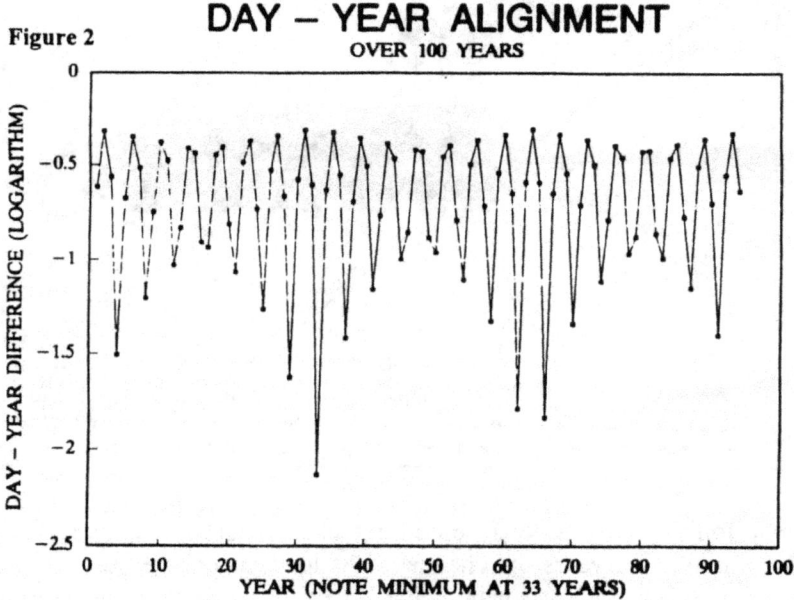

Figure 2

The vertical axis indicates the error from 364.242 days plotted on a logarithmic scale. The base line, 2.5, represents the minimum deviation. The horizontal axis indicates years from 0 to 100. This clearly indicates the thirty three year pattern. ***Diagram courtesy of Nick Kollerstrom.***

However one interprets the data, this presents a possible astronomical source to the use of the number thirty three in heroic myths. Thus when Christ's resurrection occurs at age thirty-three, witnessed as it was at sunrise, we may be faced with a sophisticated astronomical/calendar metaphor. Even the rolling away of the stone to reveal the resurrected saviour may plausibly be argued to represent the emergence of the Sun from behind a stone marker, inaugurating another thirty three year cycle.[13]

Conclusion
Contemporary archaeology has established the astronomical sophistication of the megalith builders to a previously unsuspected level. So far the arguments, naturally enough, have rested primarily on the archaeological record. However, a second line of argument may be derived from the mythical record, even though we have to account for the problem that the written sources are necessarily of a much later date than the stone remains. It is clear that myth may provide evidence of ancient astronomical knowledge, while astronomy may also provide an additional view of ancient myths.

Postscript: The Dresden Codex
Interestingly, we can find the solar cycle of thirty-three years within other cultures, such as the Mayan. I draw attention to this as additional evidence that thirty three-year cycles may have been apparent to early astronomers. The Dresden Codex, a collection of divinatory almanacs mostly tied to the 260 day divinatory cycle, contains eclipse timings compiled by the Maya and runs for almost 33 years.[14] This tabular codex abruptly finishes after 32 years and 270 days and the reason for its length is understood to be connected with the fact that 46 sacred periods of 260 days tally with this period.[15] However, it is also evident that after this time frame, which corresponds to 405 lunations, the Sun will meet the lunar node axis and produce an eclipse. In other words, 405 lunations is an eclipse cycle; after 32 years and 270 days, the nodes will have made one and three quarter revolutions whilst the Sun has made 32 and three quarter revolutions of the ecliptic, the result being that the Sun meets one of the nodes, resulting in an eclipse.[16] As the Dresden Codex is in part a document about eclipses, there was thus no real need to continue the tabulation to 33 years in order to complete the niceties of the evidently known solar cycle of thirty-three years, 405 lunations is the last eclipse possible within this cycle. It is therefore reasonable to infer that the codex recognises the importance of the 33 year solar cycle and that the Maya were familiar with it.

Table 1.

Important Solar Returns behind a horizon alignment			
Number of years	Days	Time Difference from whole number	Angular error from original solar observation
4	1,460.968	45 minutes	1 min. 30 sec.
21	7,670.086	124 minutes	3 min. 42 sec.
33	**12,052.992**	**10.7 minutes**	**0 min. 18 sec.**
62	22,645.016	23.53 minutes	0 min. 36 sec.

The Tropical Solar Year is 365.242199 days in length. Multiply this by whole numbers (of years) and look for products where the fractional part of the result tends towards zero or one. There are several contenders, shown above. Consecutive years contain an angular error of one quarter (15 minutes) of a degree. The Daily angular sunrise change along the horizon in Southern Britain at the equinox is over 0.7 degree. This is considerably more than one solar disc diameter (about 0.6 degree).

References

1. Alexander Thom, *Megalithic Sites in Britain*, (Oxford, Oxford University Press, 1967, pp 59, 165. Thom reckoned that good observation can detect angular changes as small as 2.5 minutes of arc (correspondence from Archie Thom to the author, 1994). John E.Wood, *Sun, Moon and Standing Stones*, (Oxford, Oxford University Press, 1980) wrote that 'the Temple Wood observatory shows inherent accuracies of declination measurement to around one hundredth of a degree'.
2. Wood, *Sun, Moon*, pp 36-56
3. The best source for the early myths is the Royal Irish Academy's edition *of The Book of Leinster*, 1880, a facsimile of the *Leabar na Nuachonghbhala* (also sometimes known as The book of Glendalough). This contains the earliest known version of the *Leabhar Gabhala*, the 'book of invasions', the primary source for the stories of the Tuatha de Danaan.. The most reliable version of this work is *Lebor Gabala Erenn: Book of the Taking of Ireland,* ed. R.A.S.MacAlister, Irish Text Society, 5 vols, 1938, 1939, 1940, 1941, 1954.
4. N.L.Thomas, *Irish Symbols of 3,500 BC* (Mercia Press, Cork, 1988), p 83.
5. Thomas, *Irish Symbols*, p 76. For The White Book of Rhydderch *see Llyfr Gwyn Rhydderch (The White Book of Rhydderch),* with an introduction by T.M.Jones, (University of Wales Press, Cardiff, 1973, reprint of the 1907 edition edited by J.Gwenogvryn Evans) The original manuscript is National Library of Wales MSS Peniarth 4. Also see Rees, Alwyn and Brinley, *Celtic Heritage* (Thames and Hudson, London, 1961), pp 200 f., 318 ff., 338. Although *The White Book of Rhydderch* is dated to 1300-25, like other similar texts, it is widely believed to represent the written account of a much earlier oral tradition.

6. Luke 3.23, Acts 2, record that Christ's ministry began at age thirty.
7. The earliest reference to the celebration of Christmas on Sol Invictus, December 25, is the Philocalian Calendar of 336. There are different formulae establishing the celebration of Easter. The standard has become the Roman version: Easter Sunday is the first Sunday after the full moon following the spring equinox. See also Henry Chadwick, *The Early Church*, Harmondsworth, Middlesex, for discussion of Christian reverence for the sun in the 1st-3rd centuries.
8. The concept of the solar hero has been particularly popularized by the Jungians. Jung wrote that 'It is not enough for the primitive to see the sun rise and set; this external observation must at the same time be a psychic happening: the sun in its course must represent the fate of a god or hero who, in the last analysis, dwells nowhere except in the soul of man'; C.G.Jung, 'Archetypes of the Collective Unconscious', *Collected Works*, Vol. 9, part 1, p 6, trans F.R.C.Hull (Routledge and Kegan Paul, London, 1959).
9. For Stonehenge see Hugh Thurston, *Early Astronomy* (New York 1994), pp 45-55.
10. Thom, *Megalithic Sites*, p 108
11. Norman Lockyer argued that simple observation was sufficient to establish these figures: 'Had ignorance led to the establishment of a year of 360 days, yet experience would have led to its rejection in a few years...If observations of the Sun at solstice or equinox had been alone made use of, the true length of the year would have been determined in a few years'. Norman Lockyer, *The Dawn of Astronomy*, (Cambridge University Press, 1894) pp 245-6, reprint MSI, 1964.
12. At Loughcrew, in Ireland, Cairn F, Stone C1, there are a set of 62 inscribed markers, whilst nearby, at Fourknocks Passage, one may count three columns of eleven chevrons, totalling 33, picked onto a stone. See N.L.Thomas, *Irish Symbols*, p 73.
13. Matthew 28:1, 'as it began to dawn, towards the first day of the week'. In first century Greek astrology the first day of the week , and the first hour of the first day, were ruled by the Sun. If we can interpret the astronomical features of the resurrection story as an allegory of the solar cycle, then Mary, as the mother, represents the origin of the process, in other words the first measurement or alignment with the stone marker thirty three years previously. To extend the allegory, the stone blocking the tomb, the entrance to the underworld, rolls away revealing the resurrected form and his entrance back into the visible world.
14. See Floyd Lounsbury's paper in the *Dictionary of Scientific Biography*, Vol 15, supplement 1, Charles Gillespie general editor (Charles Scribener's Sons, New York)
15. Evan Hadingham, *Early Man and the Cosmos,* William Heinemann (London 1983), p 223. See also Anthony Aveni, *Skywatchers of Ancient Mexico*, (University of Texas, 1980).
16. See Thurston, *Early Astronomy*, p 201.

EarlyGreek Cosmology: A Historiographical Review

Norriss S. Hetherington*

Introduction
Early Greek cosmology has attracted much attention from classicists, historians, philosophers, and scientists, with each group bringing to the subject its own interests and biases. Purportedly authoritative reconstructions and analyses of ancient Greek cosmology exist in abundance, even though no philosophical writings of the Presocratic period, circa 600 to 400 BC, have survived. The Greeks' attempt to explain celestial phenomena in natural terms and to avoid supernatural or divine intervention is a common theme linking many otherwise disparate scholarly studies. A frequent point of dispute involves the degree to which ancient ideas are to be judged in the context of modern science.

From scientist-historians, by which I mean scientists who have become historians, we have, in the words of Victor Thoren, 'modern commentaries, written fairly uniformly over the last 150 years by men uniformly possessed of more astronomical ingenuity than historical perspective or critical sense. The result is a corpus of secondary material replete with literally incredible claims, many of them mutually (and some of them self-) contradictory.'[1] At the other end of the scholarly spectrum reside classicists, producing philological rather than philosophical or historical books. As William Stahlman wrote, 'The trees are here and accurately labelled, but we never see the forest.'[2]

Between these two extremes lie a few studies of ancient Greek cosmology in its cultural context, most often focusing on the Ionian school, or Milesians, chiefly Thales (c.600 BC), Anaximander (c.610-545 BC), Anaximenes (c.546 BC) and Heraclitus (c.540-c.480 BC), and Pythagoras (b.c.570/580 BC) and the Pythagoreans.[3] It is these who chiefly concern me here. However, neither most of the accounts of the Milesians nor the Pythagoreans adequately cover all the varieties of Presocratic cosmological thought, and historians are open to the charge of over-simplification, that 'for long enough we have thought of early Greek philosophy as a tennis match between Ionians and Italians, with all the Greeks in the middle gaping dumbly up as the ball flew to and fro above their heads'.[4]

* University of California, Berkeley,

Sources
One of the most useful and convenient collections of the raw materials for reconstructions and analyses is contained in *The Presocratic Philosophers* by Kirk, Raven, and Schofield. Their work includes extant fragments, mainly a few quotations from Presocratic works that have survived in books written after 400 BC, with the Greek original and English translation, and commentaries. Their principal sources are testimonia, comments in the writings, such as have survived, of Plato, Aristotle, and Theophrastus, written shortly after the Presocratic period; and the doxographical tradition, consisting of summaries of the works of Plato, Aristotle, and Theophrastus, and summaries of summaries, the primary source for the summarizers being the multi-volume history of early philosophy by Theophrastus.[5]

Yet none of the surviving sources is above question. The fragments probably are the most reliable, or least unreliable, but there is no surviving original against which to check them. The testimonia have an additional uncertainty. Aristotle gives serious attention to the earlier philosophers, but his judgment and his corresponding analysis and description of earlier philosophy may well have been distorted by his own belief in the importance of the material nature of the world. He desired 'to find predictions of his own conclusions in the works of his predecessors'.[6] Plato, in contrast, offers only casual remarks on his predecessors. Just how little has survived is illustrated by the example of Theophrastus. Of the approximately eighteen books he wrote, at most a handful have survived, and the doxographical tradition consists primarily of summaries of his books and summaries of summaries.

Classicists are virtually unanimous in doubting the reliability of the surviving sources, though with differing degrees of forcefulness. In *Early Greek Astronomy to Aristotle*, D. R. Dicks' argumentative scepticism left readers with a decidedly negative aftertaste.[7] He was even more polemical and irritating in journal articles and in 1966 he wrote that 'The literature is now full of references to the scientific achievements (so-called) of the Presocratics, and the earlier the figure (and consequently the less information of reliable authenticity we have of him) the more enthusiastically do scholars enlarge his scientific knowledge'...[8] Thales is the earliest such figure and Dicks considered that 'Inevitably there accumulated round the name of Thales, as that of Pythagoras (the two often being confused), a number of anecdotes of varying degrees of plausibility and of no historical worth whatsoever'.[9]

Speculation or Science?

Having dispensed with modern scholars, Dicks turned on their ancient subject, the Presocratics. He argued that 'Greek astronomy was still in the pre-scientific stage. Observations of astronomical phenomena...were rough-and-ready observations, unsystematically recorded and imperfectly understood, of practical men...whose main concern was to have some sort of guide for the regular business of everyday life...Ionian speculation seems to have taken very little note of such observation (some of its wilder flights of fancy might have been avoided, if it had taken more)...Not until Ionian speculation had played itself out and it was becoming increasingly obvious that such presumptive theorising bore little or no relation to the gradually accumulating stock of observational data, did mathematical astronomy even begin to develop'.[10]

In response to Dicks' characterisation of Ionian philosophy as a speculative enterprise without a scientific future and a philosophic sideline with no impact on the development of observational science or mathematical astronomy, one critic charged that what he offered was 'essentially a Baconian or neo-Baconian view of science which admits mathematical computation together with empirical observation as the necessary characteristics of science, but which denies any role to speculative hypotheses of a strongly theoretical nature'.[11] A classicist limiting himself to Greek and Roman subjects, Dicks did not look ahead to the influence of Pythagorean number mysticism on modern science and he gave short shrift to the cosmological fantasies of the Presocratics, rejecting sweeping statements from other scholars about supposed striking similarities between patterns of thought in ancient Greek and modern science. He did, though, concede that the Pythagoreans were beginning to move away from speculative thinking. Other scholars have seen in emerging Ionian rationalism the removal almost at a single stroke of the entire mythological scaffolding of earlier, pre-scientific thought.[12]

Ionian Rationalism

Focusing on the method rather than substance of Presocratic thought avoids the difficulty that most Presocratic theories are known to be false. To put it bluntly, as Jonathan Barnes did, 'none of the Milesian theories is true: the Milesians do not compose a Greek Royal Society; and their Transactions would not make any contribution to the sum of scientific knowledge'. Further, by focusing on the rational, philosophical element within Presocratic methodology rather than the mathematical, quantifiable element, historians can avoid the difficulty that 'none of the Milesians aspired to the sort of precision we require in a scientific theory: their views are incurably vague; and underlying this vagueness is a complete innocence of the delights of measurement and quantification'.[13]

An emphasis on rationalism as a key characteristic of Presocratic cosmology also fits nicely with the 'Greek miracle' view of ancient history. Simply stated this holds that in the beginning there were 'charming but childish Egyptians and Sumerians with their weird and fantastic notions about the cow-goddess in the sky, the sweet waters under the earth, and so on, and then along came the Greeks who were adult rational people like ourselves'.[14]

Francis Cornford, a historian of ancient philosophy, did much to establish the belief in the miracle of Presocratic rationalism. In *From Religion to Philosophy*, first published in 1912 but still popular, he commented on the absence from Milesian philosophy of astrological superstition, magical powers, and mythical cosmogony. Less noted by readers is his conclusion that the advent of the new rational spirit was not a sudden and complete breach with the old, and that there remained a thread of continuity from science back to the supernatural world of the gods.[15]

Cornford's thesis has not been superseded, even if part of it, no doubt compatible with prevailing theories of race in England early in the twentieth century, now risks being found less than perfectly politically correct. 'The scientific tendency is Ionian in origin: it takes its rise among that race which had shaped Homeric theology, and it is the characteristic product of the same racial temperament.'[16] Cornford's Presocratics also aspired to be modern scientists. 'The aim of science...triumphantly achieved...succeeded in reducing physics to a perfectly clear, conceptual model, such as science desires...'[17] But the Presocratics could not have known what lay in the future. 'They were not trying to give a scientific system, since no one yet had told them what "science" ought to be.'[18]

By the 1930s Cornford had forged general agreement on the proposition that with Thales what we call western science first appeared in the world. He wrote that 'The intelligence became disinterested and now felt free to voyage on seas of thought strange to minds bent on immediate problems of action. Reason sought and found truth that was universal, but might, or might not, be useful for the exigencies of life...Science begins when it is understood that the universe is a natural whole, with unchanging ways of its own - ways that may be ascertainable by human reason, but are beyond the control of human action'.[19] With the rise of science, he thought, there occurred a corresponding demise of magic and mythology, themselves pre-scientific practices designed to bring supernatural forces under some measure of control.

Later, Cornford had second thoughts. All his life he was a dissenter, and ultimately he dissented against the proposition he had done so much to establish, that Ionian natural philosophers were scientific. Now he noted that their dogmatic pronouncements easily could have been upset by careful observation or the simplest experiment, but they had no empirical theory of knowledge to govern their speculations.[20] Published only posthumously, Cornford's second thoughts still have not entirely overtaken his first. A new generation of historians is more sensitive to the unscientific nature of Presocratic science, but some writers still single out rationalism, seemingly unaware both that they are echoing Cornford and that he has retracted much of the foundation upon which their derivative accounts are constructed. It is not inconsistent, however, to maintain that the Presocratics had a scientific world-view even though they lacked the experimental method.[21]

Benjamin Farrington was prominent amongst those who emphasised the rational in Presocractic cosmology. In his *Science in Antiquity*, he traced the development of ancient science in close relation to the history of philosophy. For the Milesians, he argued that Thales' importance lay in his being the first person known to have offered a general explanation of nature without invoking the aid of any outside power. He concluded that Anaximander's brilliant advance was toward a more abstract conception of nature; no longer was the underlying substance of the material world a visible, tangible state of matter, such as water, but the lowest common denominator of all sensible things arrived at by a process of abstraction.

Farrington placed the Pythagoreans, on the other hand, in a context of a spiritual revival brought about by the menace of the Persian advance, arguing that Pythagorean mathematics was primarily a religious exercise.[22] Here he was followed by other scholars, such as June Goodfield and Stephen Toulmin, who argued that 'Pythagoras, it is clear, was not so much the leader of a scientific research team, or the principal of an educational establishment, as (in modern terms) the guru of an Indian ashram'.[23] Earlier Bertrand Russell had found modern parallels for

Pythagoras when he described him as 'a combination of Einstein and Mrs. Eddy'.[24]

Science and Society
Farrington forged a very different thesis in *Greek Science*, exploring connections with practical life, with techniques, and with the economic basis and productivity of Greek society. A Marxist, he was interested in the effect of class interests in determining early Greek philosophical opinion. Egyptian and Babylonian cosmogonies, known to Thales, had embodied the idea of water in the beginning, probably because the land in both countries had been won in a desperate struggle with nature by draining swamps. Thales, leaving out only the god who had let dry land be, still formed everything out of water. Farrington seemingly chose to ignore Aristotle's speculation that Thales may have arrived at his supposition from seeing the nurture of all things to be moist. Continuing his emphasis on the practical, Farrington speculated that Heraclitus chose fire as the first principle perhaps because it was the active agent which produced change in so many technical and natural processes. In marked contrast was Pythagorean society, in which contempt for manual labour kept pace with the growth of slavery and technical processes of production became more shameful, fit only for slaves. How fortunate and acceptable that the secret constitution of things was revealed not to those who manipulated nature but to the thinkers. This, Farrington speculated, marked the separation of philosophy from the techniques of production.[25]

The possible links between Presocratic cosmology and its social setting are not limited to Farrington's imagination. Jean Pierre Vernant set out to explain the change from arithmetical Babylonian astronomy to geometrical Greek cosmology by arguing from the general premise that social change preceded philosophical. Thus the rationalisation of science and cosmology followed the secularisation and rationalisation of political administration. He related this process to the reorganisation of social space within the city and the appearance of the open central public space, the agora, in Ionian and Greek cities. Consequently, he argued that cosmological space was reorganised when Anaximander placed the earth in the centre of the universe.[26] Such speculation seems to have quickly exhausted the few facts we think we know regarding the Presocratics.

While Farrington, the Marxist, argued that science and cosmology are derived from social structures and needs, others have searched for ways in which science and cosmology influenced society. The historian of science, Richard Olson, for instance, has sought out instances of the extension and application of scientific attitudes and modes of thought beyond the domain of natural phenomena to a wide range of cultural issues that involve human interactions and value structures. He concludes that the rise of Presocratic science and the intrusion of its 'attitudes and ideas into a collapsing intellectual structure accelerated the downfall of traditional beliefs, and was decisive in shaping and forming the religious, ideological, and moral traditions that replaced those grounded in Homer'.[27] The general thesis seems plausible enough, especially for modern times in which science and technology play an increasingly larger role in our lives, but a detailed and convincing articulation of the theme for the Presocratic period remains to be done.

Koestler's Sleepwalkers
Another innovative approach to Presocratic science flowed from the pen of the novelist Arthur Koestler into his book *The Sleepwalkers: A History of Man's Changing Vision of the Universe*. Koestler was interested in the psychological process of discovery and in the process that initially blinds a person towards truth which, once perceived, is regarded as heartbreakingly obvious.[28] He examined the unconscious biases and philosophical and political prejudices of astronomers and scientists more than a decade before the physicist and historian of science Gerald Holton coined the term *themata* to describe the underlying beliefs, values, and world views that lie behind the quasi-aesthetic choices that scientists make and which guide their leaps across the chasm between experience and basic principle.[29] No branch of science, Koestler asserted, whether ancient or modern, could claim freedom from metaphysical bias of one kind or another. Although the progress of science generally was regarded as a clean, rational advance along a straight, ascending line, in fact it zigged and zagged, so Koestler argued, nearly two decades before Thomas Kuhn questioned the common notion of scientific progress.[30] He saw that the history of cosmic theories, in particular, was a history of collective obsessions and controlled schizophrenias, more of a sleepwalker's performance than of an electronic brain.

Then, in Kepler's unfolding story, came Pythagoras, whose influence on the ideas of the human race was all-encompassing, uniting religion and science, mathematics and music. He took the first steps toward the mathematisation of human experience, the beginning of science. His emphasis was on form, proportion, and pattern, on the relation rather than on the relata. The Pythagorean dream that musical harmony governed the motion of the stars, though, more a dream dreamt through a mystic's ear than a working hypothesis, more a poetic conceit than a scientific concept, retained its mysterious impact, reverberating through the centuries and calling forth responses from the depth of the unconscious mind. In the sixteenth century Kepler, enamoured of the Pythagorean dream, used its foundation of fantasy to build, by equally unsound reasoning, the solid edifice of modern astronomy.

Scientific Highlights
Koestler attracted few followers, and interest in Presocratic cosmology moved back to the ideas themselves and to the argument that they represented increasing rationality on the road to modern science. Marshall Clagett, one of the first professional historians of science emerging from university studies in the United States after World War II, insisted in his *Greek Science in Antiquity* that the tone of much of Presocratic philosophy was rational, critical, often secular, and non-mythological. He argued that the critical spirit that emerged from this period was of great significance for the subsequent growth of science, especially the emergence of a theoretical and abstract science, in which sets of empirical rules were replaced by more generalised ones. Clagett did admit that the schemes of Thales and his successors originated in analogies and patently insufficient observational data. The Pythagoreans, on the other hand, used mathematics to deepen the ties between their theoretical explanation of nature on the one hand and their experience of nature on the other.[31] Clagett showed enthusiasm, sympathy, and understanding, and his book has yet to be displaced. It has, though, been characterised as the last of the old-style general handbooks, concentrating on science separately from its philosophical background, as a history of scientific highlights, rather than an attempt to understand both ancient science and the society that produced it.[32]

Source books constitute another category of scholarly text, and for Presocratic science there is *A Source Book in Greek Science*. The editors, Morris Cohen and I. E. Drabkin, a philosopher and a classicist, realised that it was an error to study the past exclusively from the point of view of current conceptions, judging ancient science according to modern criteria, but they also were concerned to discriminate between genuine science and folklore.[33] The resulting book has been criticised for looking at

18 Early Greek Cosmology: A Historiographical Review

ancient science through modern quantitative spectacles, concentrating on only the highest and most successful examples, which happen to be mathematics, astronomy, and mathematical geography. The editors included only what they regarded as scientific material and omitted any reference to philosophical speculation. In any future source book more attention should be given to the intellectual background and how the ancients organised and systematised their own thinking about nature. The case that this should be so has already been convincingly made.[34]

Cohen and Drabkin's failure to consider the broader context of the development of quantifiable science is hardly new, and is characteristic of many previous histories. Thomas Kuhn, a critic of such work, described its methods. He wrote that it sought 'to clarify and deepen an understanding of contemporary scientific methods or concepts by displaying their evolution. Committed to such goals, the historian characteristically chose a single established science or branch of science - one whose status as sound knowledge could scarcely be doubted - and described when, where, and how the elements that in his day constituted its subject matter and presumptive method had come into being. Observations, laws, or theories which contemporary science had set aside as error or irrelevancy were seldom considered unless they pointed a methodological moral or explained a prolonged period of apparent sterility'. In Kuhn's words the scientist-historian viewed 'the development of science as a quasi-mechanical march of the intellect, the successive surrender of nature's secrets to sound methods skilfully deployed'. Only gradually have historians of science come 'to see their subject matter as something different from a chronology of accumulating positive achievement in a technical specialty defined by hindsight'.[35]

Precursor to Modern Science
One of the most positive appraisals of Milesian philosophy vis-à-vis modern science was written in the 1950s by S.Sambursky, a physicist. In *The Physical World of the Greeks* he purported to find striking similarities between patterns of thought in ancient Greek and modern science, and he presented 'noteworthy examples of the scientific approach that in the sixth century opened up a new era in the history of systematic thought...the teaching of the Milesian philosophers, which is remarkable for its rationalism'.[36] Sambursky was less enthusiastic about the Pythagoreans, though he did concede that their application of mathematics to basic physical phenomena conformed with correct

modern method. Despite the similarity of formal scientific approach, however, Sambursky claimed an essential difference: the Pythagoreans extrapolated humanity into the cosmos, while modern science attempts to project mathematical and physical laws into man.[37]

Sambursky came to his study with a background in science and emphasised the emergence of modern science. Jacob Bronowski, on the other hand, came as a mathematician and, not surprisingly, focused his attention and his film 'The Music of the Spheres' (in *The Ascent of Man* series) on Pythagoras's search for a basic relation between mathematics and phenomena of nature. In his view progress in astronomy and physics followed from their amenability to mathematical treatment, and the laws of nature have been made of numbers since Pythagoras said number was the language of nature.[38]

Even more positive regarding the Pythagorean contribution to human advance were Olaf Pedersen, a historian of science, and his co-author Mogens Pihl, a physicist. *In Early Physics and Astronomy: A Historical Introduction,* they wrote that 'there is a mathematical structure behind the visible universe; the description of nature must therefore be expressed in terms of mathematics. From now on [after Pythagoras], this connection between physics and mathematics takes a progressively stronger hold upon the minds of natural philosophers, and must be thought of as the most important contribution to the advancement of science made by the Pythagoreans. It retained its fascination, and its inspiration to scientists persisted even after the specific Pythagorean doctrines had been abandoned as naïve or as obscure manifestations of an arbitrary number mysticism'.[39] This book was intended as an introductory textbook, and certain choices had to be made. Still, as one reviewer asked: 'Is it best to present the many physical concepts in a manner that is readily intelligible to the modern reader even if it means mathematising what was often rendered verbally and wrenching out of context ideas that may have been submerged in philosophical, metaphysical, and even theological discussion? And is it justifiable to concentrate on those aspects of ancient and medieval physical thought that adumbrated or heralded ideas and concepts that would prove significant in the scientific revolution, while ignoring by far the largest portion of early physical thought which might strike the modern reader as crude and irrelevant?'[40]

Rational Debate

Enthusiasm was expressed for the Milesians by the classicist G.E.R.Lloyd, though not because of any purported similarities with modern science. In *Early Greek Science: Thales to Aristotle*, Lloyd finds two major achievements of Milesian philosophy: the rejection of supernatural explanations of natural phenomena and the institution of the practice of rational criticism and debate. He saw that dogmatic though Presocratic philosophers were in presenting their answers, still they tackled the same problems, investigated the same natural phenomena, and were aware of the need to examine and assess their opponents' theories. Lloyd attributes this practice of debate to political conditions in Greece and an extension of the customs of political debate to scientific inquiry.[41] 'But', he wrote, 'while philosophy and science did not involve a different mentality or a new logic, they may be represented as originating from the exceptional exposure, criticism and rejection of deep-seated beliefs...So far as an additional distinctively Greek factor is concerned, our most promising clue (to put it no more strongly) lies in the development of a particular social and political situation in ancient Greece, especially the experience of radical debate and confrontation in small-scale, face-to-face societies...(and) those who deployed evidence and argument were at an advantage...'[42]

Jonathan Barnes, too, in *The Presocratic Philosophers* emphasises the role of open debate in the development of cosmology and considers that 'What is significant is not that theology yielded to science or gods to natural forces, but rather that unargued fables were replaced by argued theory, that dogma gave way to reason...Few Presocratic opinions are true; fewer still are well grounded. For all that, they are, in a mild but significant sense, rational: they are characteristically supported by argument, buttressed by reasons, established upon evidence'.[43] Farrington earlier had argued from a different position in *Science and Politics in the Ancient World*. Far from political debate fostering science, he argued that scientific activity had declined in the ancient world when the struggle between science and obscurantism ultimately became a political struggle. He thought that scientific schools did save the Greeks from hierarchic petrification, but only temporarily. He drew attention to the threat Ionian philosophy posed to the institution of the state cult, and the Ionian philosopher Anaxagoras' (c.500 - c.428 BC) expulsion from Athens after his new theory of universal order posed a threat to the popular belief that celestial phenomena were controlled by the gods.[44]

Actually, Anaxagoras was indicted both for impiety and for corresponding with agents of Persia, whose subject he formerly had been.[45] Some historians choose to believe that the jury which judged him responded at least in part to the avowed charges of impiety, while other scholars elect 'to emphasise immediate political reasons for the persecution and downplay the claimed science-impiety association as an incidental rationale, unimportant in itself'.[46]

Ludwig Edelstein, a classicist, medical historian, and philosopher, was not convinced that Farrington had sufficient evidence to uphold his argument on the interaction between science and politics. 'In every respect, then,' he wrote, 'Farrington's explanation of the development of ancient science seems to be untenable. His books have done much to arouse interest in the subject. The thesis which they advocate is vitiated however by what, in my opinion, is the basic error in many of the recent evaluations of ancient science, namely, the misapplication of historical analogies. Conditions in antiquity are seen in the light of subsequent events. The conflict between science and religion, which characterised later ages, is injected into the ancient world. Progress and decay of Greek and Roman science are judged by the standards of modern science.'[47] Whatever the merits of this criticism, which Farrington had invited upon himself by citing examples from modern times to support his contention that interaction between science and politics does take place, Farrington's interpretations of ancient science have failed to attract a significant following.

Many historians are open to the charge of overemphasising in the past problems of the present. Indeed, as Richard Olson has written, there exists 'near paranoia about the whiggishness of the history of science as a discipline. We seem to agree on almost nothing but the need to avoid imposing inappropriate modern categories upon historical activities, and the need to otherwise avoid reading the present into the past. Thus, we are almost apologetic about speaking of Greek science at all...'[48]

The Pythagoreans
Lloyd, while relatively enthusiastic about the Milesians, is only lukewarm when it comes to the Pythagoreans. The two philosophies were distinguished by their religious beliefs and cosmological theories. Granted, the Pythagoreans were the first to give knowledge of nature a quantitative, mathematical foundation, and hence could be considered scientific. Yet they held not only that phenomena are expressible in

numbers, but also that things are made of numbers, this defying most modern conceptions of science. Furthermore, Lloyd concludes, 'many of the resemblances that the Pythagoreans claimed to find between things and numbers were quite fantastic and arbitrary'.[49] The Pythagoreans attracted few followers in ancient times, and Lloyd, who writes only about ancient science, rightly accords them scant attention. Those who look ahead to the Renaissance, and particularly to Kepler, though they need not attribute to the original Pythagoreans all the importance later followers achieved, cannot ignore the Pythagorean emphasis on number. At least, most cannot. James Coleman, a scientist-historian, was so distressed, however, with incorrect opinions, that when he reached Kepler in his *Early Theories of the Universe*, he could not bring himself to mention Pythagoras by name. 'Kepler, too,' he wrote, 'was a victim of the fallacious reasoning of his predecessors, but even though Kepler was often forced to many years of fruitless labour because of convictions and philosophies about the universe which he inherited, he was quick to renounce not only the erroneous arguments of predecessors but his own follies as well when this path was indicated. The clearing away of the debris enabled Kepler, with his prodigious persistence, finally to be led to the first correct description of the seemingly haphazard motions of the planets.'[50]

This perception of Kepler, fighting free of evil Pythagorean influence rather than beneficently guided by it, along with the author's focus on correctness, enabled him to see that the Pythagoreans' main contribution lay not in using mathematics to increase ties between theoretical explanation of nature on the one hand, and experience of nature on the other, as have most writers on Presocratic science, but instead in their discovery that the earth is round. Also, Coleman repeatedly found it necessary to remind his readers that the Milesians' ideas were not correct: 'That Thales was incorrect' he wrote 'is obvious in the light of the relatively vast knowledge of today...Its importance lay not in the model itself, which today is known to be incorrect, but in the fact that Anaximander was the first person to reduce the workings of the universe to a mechanical system...The "model" itself was incorrect in the light of today's knowledge, but before the facts could be established a long chain of progressively correct interpretation of astronomical discoveries had to be established.'[51]

In contrast to those scientist-historians who are perhaps more enthusiastic than erudite, Thomas L.Heath is a respected scholar whose pioneering work on Greek science remains a valuable source. In terms of emphasis and interpretation, however, his major book on early Greek astronomy falls among the older histories of science since castigated by Kuhn as chronologies of accumulating positive achievement seldom considering observations, laws, and theories which contemporary science has set aside as erroneous or irrelevant.[52] Heath set himself the stated task of 'tracing every step in the progress toward the true Copernican theory' and showing 'that Aristarchus [not Heraclides of Pontus, as Giovanni Schiaparelli had asserted] was the real originator of the Copernican hypothesis'.[53] He looked primarily at those discoveries and observations validated as scientific by modern standards: 'Thales' claim to a place in the history of scientific astronomy depends almost entirely on one achievement attributed to him, that of predicting an eclipse of the sun,'[54] while 'Anaximander boldly maintained that the earth is in the centre of the universe...'[55] The first sentence of his chapter on Anaximenes began: 'For Anaximenes of Miletus...the earth is still flat...',[56] while he described Anaxagoras as 'A great man of science (who) enriched astronomy by one epoch-making discovery. This was nothing less than the discovery of the fact that the moon does not shine by its own light but receives its light from the sun. As a result, he was able to give (though not without an admixture of error) the true explanation of eclipses.'[57] Pythagoras is credited with his eponymous theorem, with inventing the science of acoustics, his discovery regarding musical tones, and a spherical earth, but there is not even a hint that he had anything to do with some mystical philosophy regarding a relationship between mathematics and phenomena of nature. The remarkable development by later Pythagoreans, in Heath's opinion, was their abandonment of the geocentric hypothesis.[58]

Thales' Eclipse Prediction
Thales' purported eclipse prediction marks for many scientist-historians the beginning of Western astronomical science. Retrospective astronomical calculations showing a total solar eclipse on 28 May 584 BC in Northern Turkey, help confirm Herodotus' report that 'In the sixth year of the war, which they [the Medes and the Lydians] had carried on with equal fortunes, an engagement took place in which it turned out that when the battle was in progress the day suddenly became night. This alteration of the day Thales the Milesian foretold to the Ionians, setting as its limit this year in which the change actually occurred'.[59] Presumably

warring parties either took the eclipse of the sun as a sign to cease fighting, or they were eager for any reason to cease and found the eclipse a convenient excuse. Most historical discussion has centred not on the credibility of the tradition itself, but on what methods Thales could have used to predict the solar eclipse. Willy Hartner has argued that Thales could have predicted an eclipse before the end of 583 BC from a study of the periodic recurrence of solar eclipses, and then taken credit for a different eclipse occurring slightly earlier.[60]

Thales' prediction of the solar eclipse of 584 BC may, however, be more myth than historic truth and as Alden Mosshammer has pointed out, 'As modern research in the history of ancient science and mathematics has advanced, confidence in Thales' ability to predict a solar eclipse has receded'.[61] Weighing in with the most caustic damnation of the credulity of his naive colleagues is Dicks, who found their conclusions 'totally at variance with the available evidence...of Thales' alleged prediction of a solar eclipse. In a desperate attempt to vindicate the historicity of this prediction, [the scholar] spins a web of inferential reasoning, based on wholly improbable suppositions...presupposing not only accurate observations, but also the concept of the ecliptic...the assumption that such comparatively advanced astronomical knowledge was possible in the sixth century BC is ludicrous; as we have seen, all the indications are...that such a stage was not reached until at least 150 years later'.[62]

Fresh Views
Providing a welcome contrast to the older-type histories is Stephen Toulmin and June Goodfield's *The Fabric of the Heavens*. He was trained as a physicist but later became a historian of science, and she was trained by him as a historian of science. In a series of books on the development of scientific thought, they set out 'to illustrate and document the manner in which our chief scientific ideas have been formed'.[63] This beginning could all too easily have led to yet another chronology of accumulating positive achievement, but they realised that 'to understand fully the scientific traditions which we have inherited, it is not enough to discover what our predecessors believed and leave it at that: we must try to see the world through their untutored eyes, recognize the problems which faced them, and so find out for ourselves why it was that their ideas were so different from our own...Different situations gave rise in earlier times to different practical demands; different practical demands posed different intellectual problems; and the solution of these

problems called for systems of ideas which in some respects are not even comparable with our own'.[64] In other words the Presocratics did have some ideas which are now judged correct, but they did not elaborate, test them, or prove them. The union of theory and practice characteristic of modern science came later. Presocratic science was purely an intellectual enterprise undertaken with no technological end in view. For wild generalisation or unsound theorising or incautious analogy there was no potential penalty to pay in bridges collapsed or lives lost, and hence also no shackles on originality and imagination.[65] On the Pythagoreans, Toulmin and Goodfield argued that 'the most grandiose ambition they conceived was to explain all the properties of nature in arithmetical terms alone', and their 'belief that the distances of the planets from the centre of their orbits fit a simple "harmonious" mathematical law was the life-long conviction of Kepler, two thousand years later, and inspired the whole course of his astronomical researches'.[66]

The intellectual nature of Presocratic science and the separation of theory from practice are also themes in a joint appraisal of the Pythagoreans by Bernard R.Goldstein and Alan C.Bowen, a historian of science and a classicist. They wrote that 'The Pythagoreans regarded the explanation of the heavenly motions in terms of these ratios as knowledge of the speeds, risings, and settings of the celestial bodies; and Plato called it astronomy. But, though such speculation did relate celestial movement and number, it would be wrong to see in this any attempt at precise measurement of what is observed. The explanandum in these theories is not so much a physical phenomenon as the ethical and aesthetic order it supposedly exhibits.'[67]

A possible explanation of the Presocratic attitude toward theories, especially the apparent lack of interest in testing them, focuses on their emphasis on problems of cosmogony (how the world came into being) rather than of cosmology (the current structure and future evolution of the world), and the consequent direction of their scientific efforts to the past than to the present and the future. 'As might have been expected in an age whose central problem was cosmogony, i.e., a set of unobservable and unrepeatable phenomena, and which, moreover, lacked all magnifying devices, the need for increased factual knowledge and for testing assumptions by experience was hardly felt. The facts to be explained were supposed to be matters of common knowledge, and any endeavour to account for them was essentially like the effort to solve a riddle. A scientific hypothesis was a (more or less fortunate) guess and

26 Early Greek Cosmology: A Historiographical Review

the only criterion of its validity was its intrinsic plausibility.'[68] Lloyd, too, makes the point that much of the Presocratics' speculative effort was concentrated on astronomy, and though there might be attempts to verify theories with future observations, astronomy, strictly speaking, is not an experimental science, as it was impossible to vary or govern conditions of the objects under observation; direct experimentation was therefore impossible.[69]

Popper's Philosophy
The Presocratics may not have tested their theories, but did they discuss them? The matter of a tradition of critical discussion has been raised in a philosophical context by the philosopher of science Karl Popper. He asked wherein does the much discussed 'rationality' of the Presocratics lie? Not in any empiricism, because the Presocratics were critical and speculative rather than empirical. Yet when Popper wrote this, in the late 1950s, both traditional empiricist epistemology and traditional historiography of science were still, according to him, deeply influenced by the Baconian myth that science starts from observation and then slowly and cautiously proceeds to theories.[70] Science, according to this myth, began only when the speculative method was replaced by the observational method, when deduction was replaced by induction. For Popper, however, observations and experiments do not lead to an expansion of conjectural or hypothetical knowledge. Instead, observations and experiments play only the role of critical arguments, and their significance lies entirely in how they may be used to criticise theories.[71] From this theory of knowledge, it was but a short step for Popper to identify the modern rationalist tradition with the ancient Greek tradition of critical discussion. He identified the element of rationality in the Presocratics' thought in their attempt to know the world as the critical self-examination of their theories. Knowledge, Popper argued, proceeds by way of conjecture and refutation, and Presocratic philosophy developed through the clash of ideas in a critical debate.[72]

While pursuing his discussion of scientific methodology Popper somewhat inadvertently criticised Kirk's interpretation of Heraclitus. Kirk felt compelled to reply, upholding his interpretation of Heraclitus and also attempting to chip away at Popper's view of science.[73] Lloyd dismisses much of the squabble between Popper and Kirk as more of a difference between academic specialities than a disagreement over content. He also shows persuasively that broadening the question of scientific methodology to other fields of early Greek science produces different answers and rightly considers it a minor scandal that the debate initiated by Popper fizzled out so quickly.[74]

Following Popper, Lloyd further emphasised debate among the Presocratics in an article a few years later. 'Greek cosmology is nothing if not dialectical. And this is not an accidental or contingent feature of Greek cosmology, but of the essence of the Greek contribution.'[75] Greek cosmologists were in competition with each other for the best explanation, for the most adequate theory, and had 'an awareness of the need to examine and assess theories in the light of the grounds adduced for them...The history of early Greek cosmology is one of argument and counter-argument with a paucity of references to empirical data, and those mostly familiar ones'.[76]

Conclusion

Popper's philosophical emphasis has not won over classicists. Indeed, much of the disagreement over the nature of Presocratic cosmology can be understood in terms of the interests of different academic specialities and different assumptions about the nature of science. As Holton has observed: 'The search for answers in the history of science is itself imbued with themata...we must be prepared for the criticisms of those who are afflicted, not with our themata, but with their antithemata.'[77] And from Lloyd: 'whether or not historians make explicit their views on the philosophy of science, the history they write will inevitably incorporate judgements, on the nature of science itself, on what demarcates it from other inquiries, on scientific methodology.'[78]

With a limited amount of raw material, each new thesis quickly exhausts inherent possibilities. Interpretations of Pythagorean and Milesian cosmology and culture have little chance of becoming paradigms for the practice of what we might term 'normal' history, in analogy to Thomas Kuhn's normal science, which finds practitioners in agreement upon certain basic problems and techniques and industriously expanding and elaborating an initial idea. Here, Lloyd's work may turn out to be a happy exception to the general absence of sustainable intellectual themes in studies of early Greek cosmology. Also, Lloyd avoids the extremes of both scientist-historian and classicist; he offers historical perspective tempered by critical sense, and shows the forest as well as the trees. It is no easier to imagine means for testing speculations about Presocratic cosmology and culture than it was for the Presocratics to test their own speculations. Caught up in an intellectual speculative fever, we must be cautious lest we become so entranced that we lose our footing.

References

1. Victor E. Thoren, review of D. R. Dicks, *Early Greek Astronomy to Aristotle,* (Ithaca, New York: Cornell University Press, 1970), in Isis, 61 (1970), 541-2.
2. William D. Stahlman, review of G. S. Kirk, ed., *Heraclitus, the Cosmic Fragments* (Cambridge: Cambridge University Press, 1954), in Isis, 45 (1954), 308-9. For disagreement among the trees, see Gregory Vlastos, 'On Heraclitus', *American Journal of Philology,* 76 (1955), 337-68; reprinted in David J. Furley and R. E. Allen, eds., Studies in Presocratic Philosophy. Vol. 1, *The Beginnings of Philosophy* (London: Routledge & Kegan Paul, 1970), pp. 413-29.
3. A very brief characterisation of the two philosophies has the Milesians driven by intellectual curiosity and dissatisfaction with the old mythological models to create a systematic natural explanation for physical and celestial phenomena, but their theories tended to be untestable and dogmatic, and if evidence clashed with dogma they preferred the dogma. The Pythagoreans were more concerned with metaphysical explanations and driven more by religious imperatives than scientific ones. The Pythagoreans were characterized by Aristotle: 'the Pythagoreans, as they are called, devoted themselves to mathematics; they were the first to advance this study, and having been brought up in it they thought its principles were the principles of all things...things seemed in their whole nature to be modelled after numbers, and numbers seemed to be the first things in the whole of nature, they supposed the elements of numbers to be the elements of all things, and the whole heaven to be a musical scale and a number.' (*Metaphysics* I 5, 985b23-986a3) And Aristotle on the Milesians: 'Most of the first philosophers thought that principles in the form of matter were the only principles of all things: for the original source of all existing things, that from which a thing first comes-into-being and into which it is finally destroyed, the substance persisting but changing in its qualities, this they declare is the element and first principle of existing things, and for this reason they consider that there is no absolute coming-to-be or passing away, on the ground that such a nature is always preserved...for there must be some natural substance, either one or more than one, from which the other things come-into-being, while it is preserved. Over the number, however, and the form of this kind of principle they do not all agree; but Thales, the founder of this type of philosophy, says that it is water (and therefore declared that the earth is on water), perhaps taking this supposition from seeing the nurture of all things to be moist, and the warm itself coming-to-be from this and living by this (that from which they come-to-be being the principle of all things) - taking the supposition both from this and from the seeds of all things having a moist nature, water being the natural principle of moist things'(*Metaphysics* I 1, 983b6-27). An important book on Pythagoreanism is Walter Burkert's *Lore and Science in Ancient Pythagoreanism* (Harvard UP), 1972), which has a long section on astronomy. Also of note are the long chapter on astronomy in C.A.Huffman, *Philolaus of Croton* (Cambridge UP, 1993) and Maria Papathanassiou, 'The Influence of Pythagorean Philosophy on the Development of Mathematical

Astronomy', in K.I.Boudouris (ed.), *Pythagorean Philosophy* (Athens, 1992). A good overall study is D.J.Furley, *The Greek Cosmologists*, Cambridge UP, 1987.

4. M. L. West, 'Alcman and Pythagoras', *Classical Quarterly*, 61 (new series 17), (1967), 1-15.

5. G. S. Kirk, J. E. Raven, and M. Schofield, *The Presocratic Philosophers: A Critical History with a Selection of Texts*, 2nd. ed. (Cambridge: Cambridge University Press, 1983). Astronomical and cosmological material from this volume has been abstracted in Norriss S. Hetherington, *Ancient Astronomy and Civilization* (Tucson, Arizona: Pachart, 1987); see also 'Early Greek Cosmology', in Hetherington, ed., *Encyclopedia of Cosmology: Historical, Philosophical, and Scientific Foundations of Modern Cosmology* (New York: Garland, 1993), pp. 183-8, and 'The Presocratics', in Hetherington, ed., *Cosmology: Historical, Literary, Philosophical, Religious, and Scientific Perspectives* (New York: Garland, 1993), pp. 53-66. For entry into the voluminous literature on the Presocratics, see the introduction to bibliographic tools in the editor's supplement (pp. xvii-xxvii) and selective bibliographies to 1974 (pp. 527-542) and from 1973 to 1993 (pp. xxix-xlvii) in Alexander P. D. Mourelatos, *The Pre-Socratics: A Collection of Critical Essays*, revised ed. (Princeton, New Jersey: Princeton University Press, 1993), and Luis E. Navia, *The Presocratic Philosophers: An Annotated Bibliography* (New York: Garland, 1993). The major bibliography is L. Paquet, M. Roussel, and Y. Lafrance, *Les Présocratiques: Bibliographie analytique* (1879-1980), 2 vols. (Montreal: Bellarmin, 1988-89). Also see R.D.McKirahan, *Philosophy Before Socrates*, Hackett, 1994.

6. Kirk, *Heraclitus* (ref. 2), p. 30. See especially H. F. Cherniss, *Aristotle's Criticism of Presocratic Philosophy* (Baltimore: Johns Hopkins University Press, 1935); summarized in 'The Characteristics and Effects of Presocratic Philosophy', *Journal of the History of Ideas*, 12 (1951), 319-45; reprinted in Furley and Allen, *Studies in Presocratic Philosophy* (ref. 2), pp. 1-28. Cherniss' thesis is criticized not as incorrect, but perhaps as going rather too far, in W. K. C. Guthrie, 'Aristotle as a Historian: Some Preliminaries', *Journal of Hellenic Studies*, 77 (1957), 35-41; reprinted as 'Aristotle as a Historian', in Furley and Allen, *ibid.*, pp. 239-54. For a criticism, in turn, of Guthrie's article, see J. G. Stevenson, 'Aristotle as a Historian of Philosophy', *Journal of Hellenic Studies*, 94, (1974), 138-43. See also J. B. McDiarmid, 'Theophrastus on the Presocratic Causes', *Harvard Studies in Classical Philology*, 61 (1953), 85-156; reprinted, with abridgments, in Furley and Allen, ibid., pp. 178-238. An important new work, taking a fresh look at the Aristotelian view of the Presocratics, is Peter Kingsley, *Ancient Philosophy, Mystery and Magic*, Oxford UP, 1995.

7. Thoren, review of Dicks, *Early Greek Astronomy to Aristotle* (ref. 1).

8. D. R. Dicks, 'Solstices, Equinoxes, & the Presocratics', *Journal of Hellenic Studies*, 86 (1966), 26-40.

9. D. R. Dicks, 'Thales', *Classical Quarterly*, 53 (new series 9), (1959), 294-309.

10. Dicks, 'Solstices, Equinoxes, & the Presocratics', (ref 8).

11. Charles H. Kahn, 'On Early Greek Astronomy', *Journal of Hellenic Studies*, 90 (1970), 99-116. See also 'Some Remarks on the Origins of Greek Science and Philosophy', in Alan C. Bowen, ed., *Science and Philosophy in Classical Greece* (New York: Garland, 1991), pp. 1-10.

12. Jonathan Barnes, *The Presocratic Philosophers*, rev. ed. (London: Routledge & Kegan Paul, 1981), pp. 47-8; Heinrich Gomperz, 'Problems and Methods in Early Greek Science', *Journal of the History of Ideas*, 4 (1943), 61-76; reprinted in Daniel S. Robinson, ed., *Philosophical Studies by Heinrich Gomperz* (Boston: Christopher, 1953),

pp. 72-87, and in Philip P. Wiener and Aaron Noland, eds, *Roots of Scientific Thought: A Cultural Perspective* (New York: Basic Books, 1957), pp. 23-38.
13. Barnes, ibid., p. 48.
14. G. E. R. Lloyd, 'Greek Cosmologies', in Carmen Blacker and Michael Loewe, *Ancient Cosmologies* (London: George Allen & Unwin, 1975), pp. 198-224. Reprinted, with an introduction assessing scholarly debate on the topic and Lloyd's modifications and developments in his own position since the original publication of the article, in Lloyd, *Methods and Problems in Greek Science* (Cambridge: Cambridge University Press, 1991), pp. 141-63.
15. Francis Macdonald Cornford, *From Religion to Philosophy: A Study in the Origins of Western Speculation*, 2nd ed. (Sussex: Harvester Press, 1983), pp. v-vii. Original edition, London: Edward Arnold, 1912.
16. Ibid., p. 143.
17. Ibid., p. 144.
18. Giorgio de Santillana, *The Origins of Scientific Thought: from Anaximander to Proclus 600 B.C.-500 A.D.* (Chicago: University of Chicago Press, 1961), p. 21.
19. Francis Macdonald Cornford, *Before and After Socrates* (Cambridge: Cambridge University Press, 1932), pp. 5, 7-8.
20. F. M. Cornford, 'Was the Ionian Philosophy Scientific?' Journal of Hellenic Studies, 62 (1942), 1-7; reprinted in Furley and Allen, *Studies in Presocratic Philosophy* (ref. 2), pp. 29-41. See also Cornford, *Principium Sapientiae: The Origins of Greek Philosophical Thought* (Cambridge: Cambridge University Press, 1952).
21. Gregory Vlastos, 'Cornford's Principium Sapientiae', *Gnomon*, 27 (1955), 65-76; reprinted in Furley and Allen, *Studies in Presocratic Philosophy* (ref. 2), pp. 42-55. Also see W.A.Heidel, *The Heroic Age of Science* (Baltimore, 1933) and Robin Waterfield, *Before Eureka, The Presocratics and their Science*, Bristol Press, 1989, Ch. 9
22. Benjamin Farrington, *Science in Antiquity*, 2nd ed. (Oxford: Oxford University Press, 1969), pp. 20-1, 27-8.
23. Stephen Toulmin and June Goodfield, *The Fabric of the Heavens* (London: Hutchinson, 1961), p. 64.
24. Bertrand Russell, *A History of Western Philosophy and Its Connections with Political and Social Circumstances from the Earliest Times to the Present Day* (New York: Simon and Schuster, 1945), p. 31.
25. Benjamin Farrington, *Greek Science* (Harmondsworth: Penguin Books, 1953), pp. 36-40, 48-9.
26. Jean Pierre Vernant, *Mythe et pensee chez les Grecs* (Paris: Libraire Francois Maspere, 1965); translated as *Myth and Thought among the Greeks* (London: Routledge & Kegan Paul, 1983), pp. 181-6, 190.
27. Richard Olson, *Science Deified & Science Defied: The Historical Significance of Science in Western Culture. vol 1. From the Bronze Age to the Beginnings of the Modern Era ca. 3500 B.C. to ca A.D. 1640* (Berkeley: University of California Press, 1982), pp. 62, 72.
28. Arthur Koestler, *The Sleepwalkers: A History of Man's Changing Vision of the Universe* (New York: Grosset & Dunlap, 1959), p. 14.
29. Gerald Holton, 'Themata in Scientific Thought', *in The Scientific Imagination: Case Studies* (Cambridge: Cambridge University Press, 1978), pp. 3-24. An earlier version of this essay, followed by commentary, appeared in Holton, 'Themata in Scientific Thought', *Science,* 188 (1975), 328-34, and Robert K. Merton, 'Thematic Analysis in

Science: Notes on Holton's Concept', ibid., 335-8. On thematic analysis see also Holton, *Thematic Origins of Scientific Thought: Kepler to Einstein* (Cambridge, Massachusetts: Harvard University Press, 1973).
30. Thomas S. Kuhn, *The Structure of Scientific Revolutions*, 2nd ed., enlarged (Chicago: University of Chicago Press, 1970).
31. Marshall Clagett, *Greek Science in Antiquity* (London: Abelard-Schuman, 1955), pp. 34-35, 42, 43.
32. J. T. Vallance, 'Marshall Clagett's Greek Science in Antiquity: Thirty-five Years Later', *Isis*, 81 (1990), 713-21.
33. Morris R. Cohen and I. E. Drabkin, eds., *A Source Book in Greek Science*, 2nd ed. (Cambridge, Massachusetts: Harvard University Press, 1958), p. vii. See also Cohen, *A Dream's Journey: The Autobiography of Morris Raphael Cohen* (Boston: Beacon Press, 1949), p. 193.
34. Vallance, 'Marshall Clagett's Greek Science in Antiquity ', (ref. 32).
35. Thomas S. Kuhn, 'The History of Science', in *International Encyclopedia of the Social Sciences*, vol. 14 (New York: Crowell Collier and Macmillan, 1968), pp. 74-83; reprinted in Kuhn, *The Essential Tension: Selected Studies in Scientific Tradition and Change* (Chicago: University of Chicago Press, 1977), pp. 105-26.
36. S. Sambursky, *The Physical World of the Greeks* (London: Routledge and Kegan Paul, 1956), pp. 4-5. Translated by Merton Dagut from the Hebrew edition, *Kosmos shel ha-Yevanim* (Jerusalem: Bialik Institute, 1954).
37. Ibid., pp. 40, 42.
38. Jacob Bronowski, 'Music of the Spheres', 52-minute color film in *The Ascent of Man* series, no. 5 (BBC-TV and Time-Life Films, 1973, 16 mm. and videotape). See also Bronowski, *The Ascent of Man* (Boston: Little, Brown and Company, 1973), p. 187.
39. Olaf Pedersen and Mogens Pihl, *Early Physics and Astronomy: A Historical Introduction*, 2nd ed. (Cambridge: Cambridge University Press, 1994), pp. 17, 20.
40. Edward Grant, 'Physical Sciences before the Renaissance', *Journal for the History of Astronomy*, 7 (1976), 201-204. Grant's review was of the 1st edition of Pederson and Pihl.
41. G. E. R. Lloyd, *Early Greek Science: Thales to Aristotle* (London: Chatto & Windus, 1970), pp. 13-4.
42. G. E. R. Lloyd, *Magic, Reason, and Experience: Studies in the Origin and Development of Greek Science* (Cambridge: Cambridge University Press, 1979), pp. 265-6.
43. Barnes, *The Presocratic Philosophers* (ref. 12), p. 45.
44. Benjamin Farrington, *Science and Politics in the Ancient World* (London: George Allen & Unwin, 1939), pp, 70-1, 74-6.
45. T. W. Africa, *Science and the State in Greece and Rome* (New York: John Wiley & Sons, 1968), p. 39.
46. Richard Olson, 'Science, Scientism and Anti-Science in Hellenic Athens: A New Whig Interpretation', *History of Science*, 14 (1978), 179-99; Olson, *Science Deified & Science Defied* (ref. 27), pp. 79-82.
47. Ludwig Edelstein, 'Recent Trends in the Interpretation of Ancient Science', *Journal of the History of Ideas*, 1952, 13:573-604; reprinted in Wiener and Noland, eds., *Roots of Scientific Thought* (ref. 12), pp. 90-121, and in Owsei Temkin and C. Lilian Temkin, eds., *Ancient Medicine: Selected Papers of Ludwig Edelstein* (Baltimore: Johns Hopkins Press, 1967), pp. 401-39.

48. Richard Olson, 'Science, Scientism and Anti-Science', p 179-199. On whiggism, priggism, presentism, contextualism, and anti-antiwhiggism, see Stephen G. Brush, 'Scientists as Historians', *Osiris*, 10 (1995), 215-231.
49. Lloyd, *Early Greek Science* (ref. 41), pp. 24-6.
50. James A. Coleman, *Early Theories of the Universe* (New York: New American Library, 1967), p. 106.
51. Ibid., pp. 18, 21, 22.
52. Kuhn, 'The History of Science', (ref. 35).
53. Thomas Heath, *Aristarchus of Samos: The Ancient Copernicus. A History of Greek Astronomy to Aristarchus together with Aristarchus's Treatise on the Sizes and Distances of the Sun and Moon. A New Greek Text with Translation and Notes* (Oxford: Clarendon Press, 1913), p. iv. See also Heath, *Greek Astronomy* (London: J. M. Dent & Sons, 1932).
54. Heath, *Aristarchus of Samos*, p. 13.
55. Ibid., p. 24.
56. Ibid., p. 40.
57. Ibid., p. 78.
58. Ibid., pp. 46, 48, 94.
59. Herodotus, I, 74. See Kirk, Raven, and Schofield, *The Presocratic Philosophers* (ref. 5), pp. 81-2.
60. Willy Hartner, 'Eclipse Periods and Thales' Prediction of a Solar Eclipse - Historic Truth and Modern Myth', *Centaurus*, 14 (1969), 60-71.
61. Alden A. Mosshammer, 'Thales' Eclipse', *Transactions of the American Philological Association*, 111 (1981), 145-55. See also Otto Neugebauer, *Exact Sciences in Antiquity*, 2nd ed. (Providence, Rhode Island: Brown University Press, 1957), pp. 142-3.
62. Dicks, *Early Greek Astronomy* (ref. 1), p. 174.
63. Toulmin and Goodfield, *The Fabric of the Heavens* (ref. 23), p. 15.
64. Ibid., p. 16.
65. Ibid., pp. 68-9.
66. Ibid., pp. 79, 82.
67. Bernard R. Goldstein and Alan C. Bowen, 'A New View of Early Greek Astronomy', *Isis*, 74 (1983), 330-40; reprinted in Goldstein, *Theory and Observation in Ancient and Medieval Astronomy* (London: Variorum Reprints, 1985), pp. 1-11.
68. Gomperz, 'Problems and Methods in Early Greek Science', (ref. 12), pp. 31-2.
69. G. E. R. Lloyd, 'Experiment in Early Greek Philosophy and Medicine', *Proceedings of the Cambridge Philological Society*, 190 (new series 10), (1964), 50-72; reprinted, with an introduction assessing scholarly debate on the topic and Lloyd's modifications and developments in his own position since the original publication of the article, in Lloyd, *Methods and Problems in Greek Science* (ref. 44), pp. 70-99.
70. Karl R. Popper, 'Back to the Presocratics', Procedings of the Aristotelian Society, 59 (1958-1959), 1-24; reprinted, with additions, in Popper, *Conjectures and Refutations: The Growth of Scientific Knowledge* (London: Routledge and Kegan Paul, 1963), pp. 136-65, and in Furley and Allen, *Studies in Presocratic Philosophy* (ref. 2), pp. 130-153.
71. Ibid., on p. 151.
72. Ibid., on pp. 148-152.
73. G. S. Kirk, 'Popper on Science and the Presocratics', Mind, 69 (1960), 318-39; reprinted in Furley and Allen, *Studies in Presocratic Philosophy*, (ref. 2), pp. 154-77.
74. G. E. R. Lloyd, 'Popper versus Kirk: a Controversy in the Interpretation of Greek Science', *British Journal for the Philosophy of Science*, 18 (19670, 21-38; reprinted, with an

introduction assessing scholarly debate on the topic and Lloyd's modifications and developments in his own position since the original publication of the article, in Lloyd, *Methods and Problems in Greek Science* (ref. 14), pp. 100-20, esp, p 105.
75. Lloyd, 'Greek Cosmologies', (ref. 14).
76. Ibid., pp. 209, 218-9.
77. Holton, 'Themata in Scientific Thought', (ref. 29).
78. Lloyd, 'Popper versus Kirk', (ref. 74), p. 100.

Changes in Celestial Journey Literature: 1400-1650

Alan S. Weber*

Introduction

This study investigates an important historical phase in the curiously hybrid genre of the celestial journey narrative which has produced not only important scientific texts, such as Macrobius's *Somnium Scipionis*, but also some of Western Europe's finest poems, including Dante's *Divine Comedy*. I would like to compare Christine de Pizan's *Chemin de Long Estude* of 1403, which describes the author's celestial journey through the heavenly spheres, to another milestone in celestial voyage literature, Francis Godwin's English work *The Man in the Moone* of 1638. These two literary and historical endpoints illustrate the changes in European technical astronomy which occurred between 1400 and 1650, and also reveal the shift which occurred in the very nature of the celestial voyage genre. I will also briefly review other closely related early modern celestial voyage narratives written by Johannes Kepler and Bishop John Wilkins.

The True History of Lucian of Samosata (born circa 117 C.E.), the satirical story of the author's trip to the moon, is often cited as a seminal text in the tradition of celestial journey literature because it contains one central element common to all narratives of this type: expansiveness. I mean here expansiveness in all its senses; Lucian undertakes the ultimate journey, past the pillars of Heracles, the limit of the known classical world, and this breaking of boundaries allows him to overstep other restrictions on etiquette, mores, and good taste. This expansiveness also allows him to satirize Homer, Socrates, Heracles, Aristophanes, and Dionysus with impunity.

Similarly, the breadth of the genre allowed an early medieval textbook writer, Macrobius (circa late fourth century C.E.) to use his celestial wanderings through the planetary spheres as the pretext for a journey into the secrets of Pythagorean number symbolism and the cosmology of Plato's *Timaeus*. One of the great poems of celestial journeying, Dante's *Divine Comedy*, closely resembles both Lucian and Macrobius in its use of the movement metaphor to explore, magnify, and expatiate upon

---

* State University of New York, Binghamton

unknown physical and epistemological territory. As he physically encompassed vast spaces, Dante also explored great ethical, spiritual, scientific, and theological truths, presenting his poem as an encyclopedia of both theology and cosmology. Dante was the direct inspiration for Christine de Pizan's celestial voyage in the *Chemin de Long Estude;* she certainly drew her encyclopedic purpose from him. Her poem therefore represents an excellent mirror of the cosmological thought of her day as she interpreted it within a tradition well known to her.

Christine de Pizan's Cosmology
The life and works of Christine de Pizan (1364-c.1430) have recently received renewed attention as evidenced by the numerous translations and critical studies of her works as well as the international conferences held in her honor during the last ten years. Christine was born in Venice in 1364, the daughter of Thomas of Pizan, who later became court astrologer to Charles V of France. Thomas of Pizan [or Bologna] taught astrology at Bologna from 1345 to 1356. He is well known for the magical charm he used to expel the invading English armies from French soil by burying wax images of the English commanders at various locations throughout the French kingdom.

Christine grew up in the learned atmosphere of the court of the French King Charles V. She began to write extensively in the 1390s after the death of her husband Etienne de Castel. Along with a biography of King Charles V, she produced a large corpus of short lyrics, longer narrative poems, and a series of didactic works on the position of women in 14th Century society. She became well known to her contemporaries through her debate with Jean de Montreuil, Gonthier Col, and Jean Gerson over the alleged obscenity and misogyny of Jean de Meun's popular poem *Le Roman de la Rose.*

The recent recovery of one of Christine de Pizan's neglected works, *Le Livre du Chemin de Long Estude* (1403), has involved the steady critical appreciation of a dream vision dismissed in the last century by Gaston Paris as a mediocre text, with some potential value, however, for the history of ideas.[1] The poem opens with Christine's ruminations on her recently deceased husband and the vicissitudes of *Fortuna*. She then ponders the sources of evil and change in the world, attributing mutation to the battle of the elements: 'fire and water hate each other/ And one desires to destroy the other'.[2] This is probably an example of the type of passage which Gaston Paris believed important for the history of ideas.

From her descriptions of the heavens in the opening sections of the poem, it is abundantly clear that Christine endorses the standard Aristotelian-Ptolemaic model of the universe, so called because of its fusion of the fundamental principles of Aristotelian physics with Ptolemy's system of epicycles and geometrical determination of planetary orbits. Although the Aristotelian-Ptolemaic model remained the standard orthodoxy in western European universities during Christine's day, there were serious challenges to it throughout the Middle Ages, including such works as Bernardus de Silvestris's *Cosmographia*, which drew on Stoic, Hermetic, and Platonic thought.[3] It would be too simplistic, however, to divide medieval cosmology into Aristotelian and non-Aristotelian camps. Even astrology, which has often been portrayed in modern historiography as the antithesis of both Aristotelian cosmology and modern mathematical astronomy, was firmly rooted in the physical principles expounded in Aristotle's *De Caelo*, *De Generatione et Corruptione*, and the *Meteorologica*. As Richard Lemay observes, 'there can be little doubt that [Aristotle's works] supplied the scientific background of astrology during 2000 years after him'.[4] As we shall soon see, the European cosmology of Christine's age also eclectically adopted Stoic and Platonic ideas enriched by Arabic interpretations of Greek astronomy. Thus, European cosmology of the early modern period should not be characterized as slavish scholastic repetition of Aristotelian science, but by a richness of perspective, and Christine's interpretation of celestial science, which defines astronomy as a medium or bridge between divine and human knowledge, should not be lightly dismissed.

The greatest challenge to Aristotle's cosmological system ironically came from scholastic Christian theology itself, as evidenced by Bishop Etienne Tempier's condemnation in 1277 of 219 scholastic theses about the universe. Christine, however, does not mention any of the serious cosmological incompatibilities and tensions between Christianity and Aristotle, such as the question of the eternity of the world or the debate over the plurality of worlds.[5] Christine 'never ceases to consider Aristotle as the prince of philosophers' and liberally quotes extracts from Thomas Aquinas's commentary on Aristotle's *Metaphysics* in *Le Livre des Fais et Bonnes Meurs du Sage Roy Charles V.*[6] The Metaphysics also informs a large part of book two of *L'Avision Christine*, which may represent, as Glenda McLeod suggests, the first known vernacular commentary on Aristotle's *Metaphysics*.[7]

To return to Christine's narrative, she falls asleep during her reading and is visited by Sebille (Sybil) who announces that she will tell Christine the secrets of the universe because Christine is ready to conceive great knowledge ('apprestee a concevoir').[8] The double-entendre implicit in conceiving knowledge points to an important aspect of Christine's conception of science: she conceives of knowledge not as a body of facts brought from the outside world into the mind, but as an internal process of spiritualization, a birth of inward light.

Sebille next leads Christine to the fountain of Sapience, where all the great philosophers have drunk, including Socrates, the Cynics, Plato, Hermes, Seneca and her father Thomas of Pizan. From the fountain they proceed to the road of long study, a path well known to Christine. After whisking Christine around the various kingdoms of the world, Sebille leads her upwards to a high mountain, where she hears a clamour of Greek voices, no doubt the arguments of Stoics, Epicureans, Platonists, Skeptics, and Peripatetic philosophers all advancing their specific world views. Sebille next invites Christine to mount a subtle ladder extending from the heavens:

Light it was and portable
As if one could twist it around
And carry it without labour
Everywhere, if you wanted.
You would never be hindered or bothered by it.
It was not made of rope
Nor any kind of cord or wood.
I could not determine the material
But it was long, strong, and light.[9]

Sebille calls the curious and resplendent device 'l'eschiele de Speculacion' - the ladder of speculation - a shining entranceway into the heavens ordained for those, who like Christine, love subtlety and learning. Now begins her ascent into the celestial spheres offering her the occasion to detail the structure of the heavens for the reader.

Christine and Sebille first pass through the first heaven ('premier ciel'), which is made out of air. From there, they pass into the second heaven, the 'ether', characterized by its shining brilliance and clarity. Instead of the common Aristotelian definition of the ether as the fifth element (*quinta essentia*) comprising the stars and existing only beyond the sphere of the moon, Christine accepts the Stoic conception of the ether as the purer part of the upper air. They next continue on into the third heaven or the sphere of fire, traditionally located under the sphere of the moon in Aristotelian cosmology. They continue on into the fourth heaven, called 'Olimpe', meant to recall Mount Olympus, the seat of the pagan gods and a general epithet in Old French for sky or heaven. Christine's ladder finally ends within the fifth heaven, 'le firmament', a place of pervasive and blinding light.

It is here, in the fifth heaven, that Christine receives a lesson in both astronomy, the science of the movements of the fixed and wandering stars (planets), and in astrology, the knowledge of the powers and influences exerted by those celestial bodies:[10]

> [Sebille] showed me everything, and told me the names and powers of the planets, and she made every effort to teach me the courses of the moving stars, both the fixed and the wandering. And she told me the properties, the effects, the contraries, the powers and the influences, and their various arrangements.[11]

After her description of the physical structure of the heavens, Christine curiously admits her inability to provide the reader with any further details, because she had not learned astrology at school.[12] Undoubtedly she knew of the technical treatises in the library of her astrologer father Thomas of Pizan, who must have possessed a collection of astrological texts as well as instruments, but how far she proceeded in her studies cannot be determined with any accuracy.

Both Edgar Laird and Charity Cannon Willard have demonstrated how astrology formed one of the central concerns of the learned court, which included Thomas, assembled by le Sage Roy Charles V.[13] The court of Charles V was very much a centre of scientific learning, especially of Aristotelianism. Charles had commissioned Nicole Oresme to translate the works of Aristotle into the vernacular. Oresme (c.1320 - 1382) was a Professor of Theology at the College of Navarre in Paris. He made great contributions to mathematics and physical science, and wrote several cosmological works on astronomy (*De l'Espere*), divination *(Livre de Divinacions)*, and astrology (*Tractatus Contra Judicarios Astronomos*). The King was also a great patron of astrologers and founded the 'Collège de Maître Gervais' at the University of Paris for the study of astrology. Lynn Thorndike remarks about the court of Charles V: 'at this period wisdom and astrology were considered almost synonymous',[14] a viewpoint that constantly surfaces in Christine's works.

One source of Christine's astrological learning may have been one of the texts which went by the name of *On the Sphere*, such as Nicole Oresme's *De l'Espere* (late 14th Century).[15] Christine's description of the heavens echoes *De l'Espere* in several respects: both authors characterize the earth as a round ball as viewed from the moon. Both Christine and Oresme divide the heavens into five regions (in Oresme, three regions of the air, the sphere of fire, and the heavens). Oresme also mentions that Mount Olympus reaches into the upper air and Christine names one of her heavens 'Olimpe' in the *Chemin de Long Estude*.

Christine also makes reference in her poem to a wide variety of eclectic cosmological doctrines such as the planetary houses, the music of the spheres, and cosmic plenitude. She observes on her journey the houses of the planets ('les maisions que planetes ont'), and in which houses they are exalted ('quelles ont exaltacion').[16] She also describes the music of the spheres, the ordered movements of the heavens comprising 'l'armonie et belle chancon', a Platonic and Pythagorean idea widely disseminated in medieval literature despite Aristotle's skeptical rejection of the doctrine.[17] Her descriptions of the great number of heavenly bodies - 'la grant quantité pleniere/ Qui y est'[18] - reinforces the Aristotelian, Stoic, and later Christian idea of the plenitude of the cosmos. Scholastic Christian theology generally endorsed the idea of the Aristotelian and Stoic *plenum*, the absence of any empty space in the universe, in response to the *inane* or *kenon* (void space) of Epicurean physics which had denied divine providence and ordered causality. Void space implied a location where God was not, an affront to the creator's omnipresence. It is Christine's wonder at the plenitude of created nature which invites her to make her explorations through the celestial spheres.

Christine, however, finds that she cannot enter the Crystalline Heavens to see the nine orders of angels because of the present state of her corporeal body.[19] Sebille and Christine therefore descend to the sphere of the air, near the ethereal layer. There she meets the servants ('maigniee') of the 'intelligences haultaines,' the followers of the planets, sun, moon and other intelligences who are called Influences and Destinies. These Influences and Destinies are beings attached to every planet, intelligence, star, and heaven who serve them like household retainers.[20] Although she does not describe these beings at any length, they obviously act as mediums, messengers, and conduits between the divine powers of the heavenly bodies and material bodies on earth. In this section Christine seems to argue for complete astrological determinism, that our fates have been predestined by heavenly confluence. Yet she does remind her readers that God still rules the destinies from above:

> ...as soon as a man or a woman is born, however great, the destinies control their lives and assign them their proper end, good or evil, according to the domain of the course of the planets at the hour when the infant is born. But nevertheless God, who has given them this power, reigns above and takes care of what pleases Him.[21]

Among the Destinies, Christine sees the rebellions, treasons, destroyed towns, and tempests of the evil fortunes about to be rained down on earth by the Influences. She even boasts of some prophetic knowledge: she says she now knows what the effects of the 1401 comet will be; but these effects, which she refuses to reveal to the reader, will not unfold for another 20 years.[22] The latter part of the *Chemin de Long Estude* consists of a court held by Queen Reason in which the four estates (Sagesse, Richesse, Chivalrie, and Noblesse) debate the proper virtues pertaining to the prince. Although to modern readers the second part of the poem may seem like a mirror for magistrates, or a conduct book such as Christine's *Treasure of the City of Ladies,* tacked on to an exposition of natural philosophy, I will argue later, after a description of Francis Godwin's work, for the essential unity of the poem. The *jugement* genre, highly developed by Machaut,[23] in which a central question is debated in a real or mock court by learned advocates, dovetails perfectly with Christine's encyclopedic introductory section of the *Chemin de Long Estude* since the *jugement* allows for the airing of diverse opinions, just as the tradition of commentary on Aristotle weighed and synthesized conflicting propositions on the nature of the cosmos. As Barbara K. Altmann aptly puts it: 'what more suitable forum for intricate argument than a court-room scene, where a minimum of plot could supply a certain amount of tension as to outcome and where characters of allegorical or human nature could quite justifiably defend conflicting opinions at length?'[24]

Godwin, Kepler and Wilkins

Separated in culture and time by over 200 years, Francis Godwin's *Man in the Moone*[25] (published 1638) nevertheless stands in direct line with

Christine's *Chemin de Long Estude*; both works form part of the larger tradition of celestial journeys established by Lucian, Macrobius, and Dante.[26] Bishop Francis Godwin is best known for his *Catalogue of the Bishops of England since the first planting of Christian Religion in the Island, etc.*, for which he was awarded the bishopric of Llandaff by Queen Elizabeth.[27] He was one of a growing number of English clergymen, including Bishop John Wilkins, who took a serious interest in scientific matters and who attempted to reconcile astronomical advancements with divine writ. Godwin was writing after the detailed observational work of Tycho Brahe, later published by Johannes Kepler, and the *De Revolutionibus* (1543) of Copernicus, which had proposed a heliocentric universe and rotating earth while retaining the fundamental Ptolemaic principles of circular motion and epicycles. After Tycho's parallax measurements of the 1577 comet had situated this extraordinary cosmic event well beyond the lunar sphere - in other words a decaying, changing object had appeared in Aristotle's alleged realm of perfection - Aristotelian cosmology, already damaged by Copernicanism, began seriously to unravel.[28] Godwin was also writing after the telescopic discoveries of Galileo reported in the *Siderius Nuncius* (1611).

In Godwin's work Domingo Gonsales, a Spanish Hidalgo, narrates his remarkable life story. After killing a man in a duel, Gonsales departs on a series of sea voyages which finally land him on the island of St. Helena. He discovers huge geese there - gansas - which he trains to carry heavy loads. He invents a flying machine by tying the gansas together, eventually flies off the island and is picked up by a Spanish ship. He convinces the captain to take the birds aboard, and soon after, the Spanish fleet is defeated by the English navy. Gonsales escapes with his gansas who inexplicably fly straight upwards into the atmosphere towards the moon. On his lunar voyage he encounters pleasing shapes floating in the air who bring him delicious foods. When he reaches the moon, on September 21, 1599, he finds that the food given to him by the aerial beings has turned to 'a mingle mangle of dry leaves, of *Goats haire, Sheepe*, or *Goats-dung, Mosse*, and such like trash'.[29] He has been deceived by wicked spirits.

As Gonsales arrives on the moon, we begin to see Godwin's somewhat transparent purpose in writing this narrative - he uses Gonsales to advance contemporary cosmological doctrines. In a similar fashion, Christine's journey had allowed her to provide an eyewitness account of the true order of the heavens. First, Gonsales refutes the existence of the

Gonsales en route to the Moon,
from Godwin's *The Man in the Moone* (1638)

Aristotelian sphere of fire, a ring of elemental fire attached to the upper air. Gonsales discovers that the upper air is of the same temperature as that below: 'Who is there that hath not hitherto believed the uppermost Region of the Ayre to be extreme hot, as being next forsooth unto the naturall place of the Element of Fire. O Vanities, fansies, Dreames!'[30]

We remember in Christine's poem how she fears the increasing heat as she mounts the ladder of speculation towards the Aristotelian sphere of fire.[31] She recalls the arrogance of Icarus who flew too high and presumptuously into the fiery heavens. Godwin, on the other hand, following common practice among professional astronomers of his age, has rejected the existence of the sphere of fire.[32] By the beginning of the seventeenth century, Stoic and Hermetic monism (universally operating physical law) began to supersede Platonic and Aristotelian dualism in physics (separate physical laws for heaven and earth), a development which challenged the doctrine of the elemental spheres.

Godwin also questions the Aristotelian idea of the natural place of the elements which made the centre of the earth the point towards which the heavier elements earth and water were attracted. In Godwin's cosmological system, magnetic force, an idea previously outlined in William Gilbert's widely read *De Magnete* (1600), replaces the natural place of the earth as a centre of attraction. This modification allows for the moon to possess a centre of attraction,[33] which Gonsales discovers as he walks on the moon. The lesser magnetic force of the moon allows him to bound high into the air.

Most importantly, Gonsales watches the earth turning on its axis from his position on the moon. He must therefore conclude along with the Copernicans that the earth spins on its axis every twenty four hours from west to east. We are very far away from Christine's fixed earth which she sees from above, sitting 'like a little, round ball'.[34] Gonsales now realizes the blindness of the philosophers who posit two contrary motions for the heavens and the planets:

> *Philosophers* and *Mathematicians* I would should now confesse the wilfulnesse of their own blindnesse. They have made the world believe hitherto, that the Earth hath no motion. And to make that good, they are fain to attribute unto all and every of the celestiall bodies two motions, quite contrary each to other; whereof one is from the *East* to the *West*, to be performed in 24 hours; (that they imagine to be forced, *per raptum primi Mobilis*) the other from the West to the

East in severall proportions.[35]

Godwin and Gonsales, however, do not fully accept the Copernican hypothesis. Gonsales states:

> I will not go so far as *Copernicus*, that maketh the Sunne the Center of the Earth, and unmoveable, neither will I define any thing one way or other. Only this I say, allow the Earth his motion (which these eyes of mine can testifie to be his due) and these absurdities are quite taken away, every one having his single and proper Motion onely.[36]

Godwin's work has many affinities with Johannes Kepler's *Somnium*, a work on lunar astronomy published in 1634. With the aid of his mother, Kepler, who appears in the work as the character Duracotus, summons a demon who explains life on his home planet of Levania (the moon). Kepler uses this fiction, based on Lucian's *True History*, much in the same way as Godwin - to explicate lunar astronomy by shifting observational reference points from the earth to the moon. Thus, in his highly technical notes to the *Somnium*, Kepler explains eclipses, solstices, and lunar and terrestrial rotation as measured from the moon as a central reference point. Just as Godwin employed the fiction of the celestial journey to argue for terrestrial rotation, so Kepler states in his notes: 'here is the thesis of the whole *Dream*; that is, an argument in favour of the motion of the earth or rather a refutation of the argument, based on sense perception, against the motion of the earth'.[37]

John Wilkins' *The Discovery of A World in the Moone* was published in 1638, the same years as Godwin's *Man in the Moone* and four years after Kepler's *Somnium*. Wilkins was instrumental in founding the Royal Society and encouraged the study of astronomy at Oxford and London.[38] Unlike Godwin, Wilkins includes a full-blown assault on Aristotelianism. *The Discovery* clearly shows in what direction the genre of the celestial journey has travelled since the writing of Christine's *Chemin de Long Estude*. Wilkins altogether dispenses with the fictional narrative and presents a series of propositions about lunar and cosmic science. From the literary narrative revelatory of *scientia*, of which Christine de Pizan's work is a prime example, we have moved to the precursor of the scientific paper. As Marjorie Hope Nicolson points out: '[Wilkins's] *Discovery* is one of the first important books of modern "popular science", a work written by a man who knew the technicalities of science.'[39] Wilkins clearly sees himself working within the literary tradition of Lucian, Plutarch, Kepler, and Godwin and draws on these works, as well as Galileo's *Siderius Nuncius*, for his lunar science. Yet Wilkins also concludes that a list of numbered propositions supported by evidence, in contrast to literary narrative, can best convey scientific knowledge to a literate audience.

Wilkins fully accepts the Copernican hypothesis (heliocentrism and terrestrial rotation) and also refutes some fundamental tenets of Aristotelian physics. First, Wilkins entertains the possibility of the plurality of worlds rejected by Aristotle, an idea according to Wilkins, that 'doth not contradict any principle of reason'.[40] Along with Godwin, he rejects the orb of elemental fire.[41] Proposition number three, which clearly demonstrates his break with Aristotelianism, boldly states: 'that the heavens doe not consist of any such pure matter which can priveledge them from the like change and corruption, as these inferiour bodies are liable unto'.[42] Wilkins has swept aside the barrier of elemental matter, which Aristotle had located at the sphere of the moon. The heavens for him can no longer consist of incorruptible, immaterial ether, but must be made of something more readily accessible to reason and *experimentum*.

Not only do we see a shift in specific astronomical doctrines in European cosmology from the time of Christine de Pizan to the work of Godwin and Wilkins, but also a profound difference in how the cosmos was perceived. Many of the technical changes in cosmology can be summed up simply as the growing rejection of Aristotelian physics. The book of *Genesis* and the Hexameral treatises (works combining theological and physical speculation on the first six days of creation) stood at the heart of early medieval cosmology before the Latin translations of Arabic and Greek astronomical texts entered the West. One book, the Bible, provided a coherent theology, physics, and cosmogony. By Wilkins's day, however, theology and biblical exegesis were becoming increasingly irrelevant to both astronomy and astrology, as practitioners of these sciences, especially in astronomy, were focussing their work more on practical computation, geometry, and mathematics than on origins, causes, and metaphysical powers of the Deity. As Wilkins points out regarding extrapolating the nature of the world from divine writ: 'such...absurdities [about nature] have followed, when men looke for the grounds of Philosophie in the words of Scripture'.[43]

Conclusion

For Christine de Pizan, astrology and astronomy were merely subsciences of theology - astrology represented that speculative ladder which mounts into the stars and reveals the high majesty. Theology in Christine's *Livre de la Mutacion de Fortune* is 'that supercelestial science which comprises all other knowledge'.[44] The study of Aristotelian metaphysics is likewise only a stepping stone - or ladder - to divine knowledge. We see similar statements regarding the proper place of astronomical and astrological learning within French culture in the work of Christine's contemporary Nicole Oresme. Oresme writes:

> ...astrology has three very noble ends. The first is to have knowledge of such great matters, for to this, according to the philosophers, is human nature naturally inclined....The second end, and the chief, is that it gives great aid in the knowledge of God the Creator....The third end of astrology and the least important is to ascertain certain dispositions of this lower and corruptible nature, whether present or to come, and nothing beyond that.... [Coopland's translation].[45]

Oresme, like Christine, defines knowledge of God as the *summum bonum* of the natural philosopher. Both Wilkins and Godwin have made a complete separation of themselves from nature. Flying to the moon, an imaginative or ecstatic act in the world of Lucian, Macrobius, Dante and Christine de Pizan, in seventeenth-century writers becomes a possible material reality, the possibility of transporting physical bodies from point to point. In Godwin's day, schemes for flying to the moon were being seriously advanced by scientific thinkers. The 1640 edition of Wilkins's *The Discovery of a World in the Moone*, for example, contained an additional chapter outlining a device for such a lunar voyage.[46] Wilkins's 1648 work *Mathematical Magick: or, the Wonders that may be performed by Mechanical Geometry* also contains a description of a flying chariot.

Christine's journey, on the other hand, does not simply involve a movement from one physical locus to another. Christine's flight of speculation culminates in the knowledge of virtue: she undertakes an inward theological and spiritual journey. Stéphane Gompertz alerts us to the integrating and unifying function of Christine's epistemological journey in *Le Chemin de Long Estude*: 'the voyage [through the spheres] not only links the different regions of the universe, but also differing realms of knowledge; in this resides the journey's totalizing power'.[47] Elevated by her knowledge of the proper ethics of the prince, which she has heard debated in the Court of Reason, Christine undergoes both an inner and outer spiritualization in her travels to the heavenly spheres. Christine writes in the *Chemin de Long Estude*: 'through knowledge, the great treasure of the Understanding, better than gold, is engendered in our breasts; the fruit of such knowledge heals all wounds. Knowledge is the sun whose light illuminates the shadows of our thoughts with its fullness'.[48] The growth of inward knowledge involves building the bright, ethereal ladder of speculation in the mind's eye. Sebille suggests that Christine serve as a messenger to earth to report the debates in the court of Reason: Christine therefore truly becomes an 'influence', a spiritualized daemon or angel, transformed into the intermediate aetherial or pneumatic substance possessing the moral understanding that she will breathe into ('influer') the ears of earthly princes. Illuminated by ethical knowledge, she may also be likened to astrological rays transporting the heavenly influences to the terrestrial realm.

In Christine's conception of heaven, a strong link exists between divine knower and the known world; both are described as light and clear in her frequently employed Platonic light-as-knowledge metaphor. In the case of spiritual knowledge, the knower and the known are not distinct epistemological positions. Astronomy or astrology would not have to be studied in the heavens, since study implies a knowledge outside the individual; this separation could not occur in union with God. In heaven, the shining glorified bodies united to their souls in Christine's poem *La Prison de la Vie Humaine*, 'will find that they have all forms of knowledge, know all things perfectly, feel the infinite goodness of God'.[49]

In Christine's theory of knowledge, the long road of study requires moral and abstract thought, which then bridges the spatial and epistemological distance between the soul and heaven, the soul's final destination. As the two realms interrelate spiritually, there is no real concept of space or locus in Christine: distance between objects may be physical, but not spiritual, or vice versa. For Godwin and Wilkins, representing the triumph of the new physical, materialist astronomy in their celestial voyage narratives, the material universe is now distinct from the human soul.

References

1. Gaston de Paris, 'Chronique', *Romania* 10 (1881) 318: 'un ouvrage dont la valeur poétique est médiocre et qui n'a d'intérêt que pour l'histoire des idées et de l'instruction au XVe siècle'. Unless noted, all translations are my own.
2. Christine de Pizan, *Le Livre du Chemin de Long Estude*, ed. Robert Püschel (Paris 1887) 413-414: 'le feu et l'iave s'entreheent,/ A destruire l'un l'autre beent.'
3. See Edward Grant, *Planets, Stars and Orbs: The Medieval Cosmos, 1200-1687* (Cambridge 1994) 59: 'Medieval society's concept of the origin, structure, and operation of the world was drawn almost exclusively from the Aristotelian-Ptolemaic astronomical and cosmological tradition.' Bernardus Silvestris's *Cosmographia* has been translated by Winthrop Wetherbee, *The Cosmographia of Bernardus Silvestris: A Translation With Introduction and Notes* (New York 1973).
4. Richard Lemay, 'The True Place of Astrology in Medieval Science and Philosophy: Towards a Definition', *Astrology, Science and Society: Historical Essays*, ed. Patrick Curry (Woodbridge, Suffolk 1987) 57-78.
5. Both of these questions are intimately related to Aristotelianism. Aristotle had clearly stated the world was eternal and non-created, while the book of *Genesis* forced Christian theologians to argue for creation ex nihilo. Aristotle also denied the possibility of other worlds, arguing from his theory of the natural place of the elements. The heaviest element, earth, naturally falls towards the centre of the universe, i.e. the sphere of the earth, while light elements such as fire escape towards the heavens. Multiple worlds entailed multiple centers of attraction, in which case elements would be flying around the cosmos helter skelter. But denying the possibility of multiple created worlds placed a fundamental restriction on an omnipotent being. See Stephen J. Dick, *The Plurality of Worlds* (Cambridge 1982) and Pierre Duhem, *Medieval Cosmology: Theories of Infinity, Place, Time, Void, and the Plurality of Worlds*, ed. and trans. Roger Ariew (Chicago 1985) 431-510.
6. Marie-Josèphe Pinet, *Christine de Pizan 1364-1430: Étude Biographique et Littéraire* (Paris 1927) 423: 'Christine ne cesse de considérer Aristote comme le prince des philosophes.'
7. Christine de Pizan, *Christine's Vision*, trans. Glenda K. Mcleod, Garland Library of Medieval Literature, vol. 68, series B (New York 1993) 93 n.24.
8. *Chemin* 636-37. Jane Chance explores the question of female knowledge and its metaphors in 'Christine de Pizan as Literary Mother. Women's Authority and Subjectivity in "The Floure and the Leafe" and "The Assembly of Ladies"', *The City of Scholars: New Approaches to Christine de Pizan*, ed. Margarete Zimmermann and Dina De Rentiis (Berlin 1994) 245-259.
9. *Chemin* 1602-1610: 'Legiere estoit et portative/ Si qu'on la peust ertortillier/ Et porter sanz soy travaillier/ Par tout le monde, qui voulsist,/ Que ja n'empeschast ne nuisist,/ Non mie que de corde fust/ Ne d'autre file ne de fust;/ Ne je n'en congnois la matiere,/ Mais longue estoit, fort et legiere.'
10. Christine, following the practice of many medieval and Renaissance writers uses the terms 'astronomie' and 'astrologie' interchangeably. More precise writers followed Ptolemy's division in the *Tetrabiblos*, which circulated in the west in a widely read Latin translation entitled the *Quadripartitum*. Ptolemy clearly distinguished two cosmological sciences: the first part of the science of the stars consisted of the study of the appearance

and movements of the celestial bodies with respect to the earth (what today we would call astronomy); the second part considered the effects of the heavenly bodies on terrestrial events and human actions (what we would call astrology).

11. *Chemin* 1824-32: '[Sebille] tout me monstroit, et devisoit/ Des planetes les noms, la force,/ Et de moy enseignier s'efforce/ Les cours des estoilles mouvables/ Et des estans et des errables./ Si m'en dist les proprietez,/ L'effect, les contrarietez,/ Leurs forces et leurs influences/ Et leurs diverses ordenances.'

12. *Chemin* 1848-9: 'Car sience d'astrologie/ N'ay je pas a l'escole aprise.'

13. Charity Cannon Willard, 'Christine de Pizan: The Astrologer's Daughter' in *Mélanges à la Mémoire de Franco Simone*, vol. 1 (Genève 1980) 95-111; Charity Cannon Willard, *Christine de Pizan: Her Life and Works* (New York 1984) 17, 19-22, 97, 104; Edgar Laird, 'Astrology in the Court of Charles V of France, As Reflected in Oxford, St. John's College, MS 164' *Manuscripta* 34 (1990) 167-176.

14. Lynn Thorndike, *A History of Magic and Experimental Science*, 8 vols. (New York 1923 - 58) 3.585.

15. I am working from G.W. Coopland's paraphrase of Oresme's *De l'Espere* in *Nicole Oresme and the Astrologers: A Study of His Livre de Divinacions* (Liverpool 1952) 17-20.

16. *Chemin* 1934, 1943.

17. *Chemin* 1994. For an introduction to the history of this idea, along with excerpts from original texts, see Joscelyn Godwin's *Harmony of the Spheres: A Sourcebook of the Pythagorean Tradition in Music* (Rochester 1993).

18. *Chemin* 2010-11.

19. *Chemin* 2030-34.

20. *Chemin* 2095.

21. *Chemin* 2109-2118: '....aussi tost que l'omme naist/ Ou la femme, ja si grant n'est,/ Ceulx [les destinees] yci de sa vie ordenent/ Et sa droite fin lui assenent,/ Bonne ou male, selon les cours/ Ou les planetes ont leurs cours/A l'eure que l'enfant est né./ Mais toutefois Dieux, qui donne/ Leur a ce povoir, dessus est,/ Qui bien garde ce qui lui plaist.'

22. *Chemin* 2175-84.

23. For example, *Le Jugement du Roy de Behaigne* and *Le Jugement du Roy de Navarre*.

24. Barbara K. Altmann, 'Reopening the Case: Machaut's *Jugement* Poems as a Source in Christine de Pizan', *Reinterpreting Christine de Pizan*, ed. Earl Jeffrey Richards, Joan Williamson, Nadia Margolis, and Christine Reno (Athens 1992) p 137.

25. *The Man in the Moone: or a discourse of a Voyage thither by Domingo Gonsales The Speedy Messenger* (London 1638). All references to *The Man in the Moone* will be from the 2nd edition of 1657.

26. Lucian, *The True History*; Plutarch, *De Facie in Orbe Lunae*; Macrobius, *Somnium Scipionis;* Dante, *Divina Commedia*. Marjorie Hope Nicolson surveys some of this literature in *A World in the Moon: A Study of the Changing Attitude Toward the Moon in the Seventeenth and Eighteenth Centuries,* Smith College Studies in Modern Languages, vol. 17, no. 2 (Northampton, MA 1936).

27. *Dictionary of National Biography*, ed. Sir Leslie Stephen and Sir Sidney Lee, vol. 8 (Oxford 1917) 56 - 58.

28. *Planetary Astronomy from the Renaissance to the rise of Astrophysics, The General History of Astronomy,* ed. René Taton and Curtis Wilson, vol. 2, Part A: Tycho Brahe to Newton (Cambridge 1989) 5-7. See also J. L. E. Dryer, *A History of Astronomy from*

Thales to Kepler, rev. W. H. Stahl, ed. 2 (New York 1953) 365-71 and Clarisse Doris Hellman, *The Comet of 1577: Its Place in the History of Astronomy* (New York 1971).
29. Godwin 68.
30. Godwin 65.
31. *Chemin* 1703-22.
32. See Thorndike 6.1-2, 83, 384.
33. H.W. Lawton, 'Bishop Godwin's *Man in the Moone*', *Review of English Studies* 7, no. 25 (1931) 41.
34. *Chemin* 1699-1700: 'comme une petite pelote,/ Aussi ronde q'une balote.'
35. Godwin 58-59.
36. Godwin 60.
37. *Kepler's Somnium: The Dream, or Posthumous Work on Lunar Astronomy*, trans. Edward Rosen (Madison 1967) 82 n.96.
38. *Dictionary of National Biography*, 21.264 -7.
39. Marjorie Hope Nicolson, *Voyages to the Moon* (New York 1948) 93.
40. John Wilkins, *The Discovery of a World in the Moone (1638), A Facsimile Reproduction with an Introduction by Barbara Shapiro* (Delmar 1973) 25.
41. Wilkins 54.
42. Wilkins 44.
43. Wilkins 40.
44. Christine de Pizan, *Le Livre de la Mutacion de Fortune*, ed. Suzanne Solente, t. 2 (Paris 1957) 7309-14: 'la superceleste/ Science...Ou est compris science entiere.' See also Christine de Pizan, *Le Livre des Fais et Bonnes Meurs du Sage Roy Charles V*, ed. Suzanne Solente, vol. 2 (Paris 1936) 18, 'comme l'entencion finale de sapience ou de methaphisique soit pervenir à cognoistre le gouvernement de la cause premiere, c'est Dieu le glorieux, la cognoissance de l'ordre des esperes celestes, auxquelles cognoiscences impossible est venir, senon après astrologie; et toutefois à astrologie nul ne puit parvenir s'ançois n'est philosophe, geometre et arismetien; par quoy, comme il appert qu'en l'ordre des sciences astrologie et methaphisique sont tres haultes'.
45. Nicole Oresme, *Livre de Divinacions* in Coopland 113: 'la science du ciel a trois tres nobles fins. La premiere est avoir congnoissance de si tres belles choses car a ce est naturellement humain lignage enclin selon les philosophes.....La seconde fin et la plus principale d'astrologie est ce que elle donne grant aide a la congnoissance de Dieu le createur....La tierce fin d'astrologie et la moins principal est congnoistre aucunes disposicions de ceste basse nature corruptible presentes ou avenir et tant et nomplus....'.
46. Barbara Shapiro, 'Introduction,' Wilkins vi.
47. Stéphane Gompertz, 'La voyage allégorique chez Christine de Pisan', in *Voyage, Quete, Pelerinage Dans la Literature et la Civilization Medievales*, Senefiance No. 2, Cahiers du CUER MA (Paris 1976) 200: 'le voyage ne met pas seulement en contact les régions de l'univers mais aussi les domaines de la connaissance: c'est là que réside sa vertu totalisante.'
48. *Chemin* 5215-21: 'Par la quelle [la science]le grant tresor/ De conscience, meilleur que or,/ Est conceu en nostre courage,/ Dont le fruit tous maux assouage./ C'est le soueil par quel lumiere/ Ajourne o sa lueur pleniere/ Es tenebres de la pensee.'
49. *The Epistle of the Prison of Human Life* 60-61: 'ilz se trouveront avoir parfaite sapience, sachans toutes sciences, congnoissans toutes choses parfaitement, sentir l'infinie bonté de Dieu.'

Kepler's *Tertius Interveniens*

Ken Negus*

Tertius Interveniens, written in 1610, is one of Kepler's most powerful and passionate treatises on astrology, written as a defence of the subject against extremists on both sides, on the one hand those who would condemn astrology altogether, and on the other those who accepted everything said and done in its name, no matter how preposterous. Hence he is the 'third party intervening', as indicated by the title.

The following extract is near the centre of the book, and is otherwise 'central' as an expression of the main tenets of Kepler's thought on astrology. In this short passage, he comments incisively on the following topics: the non-material, 'spiritual' nature of astrology; geometry as the all-embracing archetype through which the messages of the sky are communicated to earth; the horoscope as indelible 'imprint' on the soul of the infant being born; astrology as 'music'; the rationale of progressions and directions (symbolic measures used in prediction); astrological genetics; transits and mundane (political and historical) astrology.

Not to be forgotten when reading Kepler's arguments concerning astrology is his major role in the history of science. Although he is remembered primarily as an astronomer he also had much to say about biology, geology, meteorology, medicine and many other areas. His philosophical thinking also suggests much that is far ahead of his time, including prefigurations of Jungian psychology (archetypes and the collective unconscious). But let him, in slightly abridged form, speak for himself:[1]

Thesis 64. All powers coming down from above are ruled according to Aristotle's teaching: namely, that inside this lower world or earthly sphere there is a spiritual nature, capable of expression through geometry. This nature is enlivened by geometrical and harmonic connections with the celestial lights, out of an inner drive of the Creator, not guided by reason, and itself is stimulated and driven for the use of their powers. Whether all plants and animals, as well as the Earth's sphere, possess this faculty, I cannot say. It is not an unbelievable thing,

---

* 175 Harrison Street, Princeon, NJ 08540, USA

for they have various faculties of this kind: in that the form in every plant knows how to put forth its adornment, gives the flower its colour, not materially, but formally, and also has a certain number of petals; nor [is it unbelievable] that the womb, and the seed that falls into it, has such a marvellous power to prepare all the body parts in appropriate form....The human being, however, with his soul and its lower powers has such an affinity with the heavens, as does the surface of the Earth, and this has been tested and proven in many ways, of which each is a noble pearl of astrology, and is not to be rejected along with [all of] astrology, but to be diligently preserved and interpreted.

Thesis 65. Above all, I might in truth flatter myself with having experienced this observation: that the human being, in the first igniting of his life, when he now lives independently for the first time, and can no longer live in the womb, receives the character and formation of the sky's whole stellar configuration, or the form of the conflux of radii on earth, and maintains it unto his grave. Afterwards this can be perceived in the formation of the face and the remaining bodily structure, as well as in the person's behaviour, habits and gestures, so that he might create, with his bodily form, corresponding attraction and charm for himself in the eyes of other people, and with his actions bring forth corresponding fortune. Then thereby (as well as from the mother's fantasies before the birth and from the rearing of the child thereafter), a great difference from other people is created, so that one person is brave, cheerful, joyful, self-confident; and another lethargic, lazy, neglectful, shy, forgetful, hesitant, and whatever other general characteristics there may be, which can be compared with configurations that are pleasant and exact, or complex and awkward, or also with the colours and motions of the planets. This character is not received into the body - for it is much too ungainly for that - but into the nature of the soul itself, which behaves like a point, so that she [the soul] might be transformed in points of the conflux of radii; and not only do these points impart reason to her, from which we human beings might be called reasonable above all other creatures, but she is to grasp in the first moment another kind of implanted reason - geometry - in the radii as well as in musical sounds [i.e. 'voices,' in the technical musical sense], without a lengthy learning process.

Thesis 66. Second, so it is with every plant, that it is on schedule when it is to ripen or blossom. This time is prescribed to it at its creation, and by external warmth and other means it is lengthened or shortened, but can never be totally altered. In like manner the human being's nature, upon entry into life, receives not only an instantaneous image of the sky, but also its movement, as it appears down here on earth for several days in a row, and in certain years derives from this moment the manner of outpouring this or that humour; these years are precisely and sharply indicated, based on the projection of the first few days [of life]. This is a truly marvellous thing, and is like an image or outflowing of the natural proportion of a day to a year. Thus this short time or 'tempus typicum' in human nature with all its parts is multiplied by 365; and all of natural life, out of this multiplication, remaining rigidly in its memory, is deducted and unwound as from a ball of yarn, so that then the whole future life, insofar as it deals with natural matters, in the course of a quarter-year is wound up and stored in a little bundle. Such a causality and natural proportion cannot, however, be applied to the profections,[2] for not the Ascendant and not the Sun, but Jupiter makes its revolution in 12 years, as the Moon does in 28 days, and accordingly the best of the profections should be assigned to the transits; the rest is useless noise. I have often harboured the thoughts that there is nothing to directions because we must reach out so far for their causality, and one cannot accommodate them any differently. But I must confess that nonetheless the causality resembles nature, because it requires a natural proportion; and that our experience is so clear, that they are not to be denied as true for the astrologers.

Thesis 67. Third: this is a curious thing that the nature [of the human being], which receives this character [of the sky], also favours its relatives by some similarities in the celestial constellations. When the mother is great with child and her time has come, then nature seeks out a day and hour for the birth that is comparable celestially with that of the mother's father or her brother.

Thesis 68. Fourth, every [human] nature knows not only its celestial character, but also every day's configurations and motions in the sky so well, that as often as a transiting planet comes into its character's Ascendant or other prominent place, especially into radical points, it [the nature] accepts it and is thereby variously affected and stimulated.

Thesis 69. Fifth, there is also the experience that every strong conjunction, by itself, without considering the relationship to a particular person, stimulates people in general (where a nation lives together in an ordered society), and makes them capable of acting as a community that is unified just as the stars then shine together. This was discussed in detail in my book *De stella serpentarii*. Thus I have seen many examples of epidemics in which the humours are stirred up more when strong stellar configurations are present (that is to say, the human natures are stimulated to drive out the humours). In like manner all these points - and many more could be cited, being from the same cause - and the possibility of one coming from the other could be proven and defended.

References

1. Translation by Ken Negus, 1997.
2. A forecasting technique in which the houses of the horoscope represent the years of life, beginning with the first house and the first year of life.

Belief in Astrology:
a social-psychological analysis

Martin Bauer[1] and John Durant[2]

Abstract

Social scientists have suggested several different hypotheses to account for the prevalence of belief in astrology among certain sections of the public in modern times. It has been proposed: (1) that as an elaborate and systematic belief system, astrology is attractive to people with intermediate levels of scientific knowledge [the superficial knowledge hypothesis]; (2) that belief in astrology reflects a kind of 'metaphysical unrest' that is to be found amongst those with a religious orientation but little or no integration into the structures of organized religion, perhaps as a result of 'social disintegration' consequent upon the collapse of community or upon social mobility [the metaphysical unrest hypothesis]; and (3) that belief in astrology is prevalent amongst those with an 'authoritarian character' [authoritarian personality hypothesis].

The paper tests these hypotheses against the results of British survey data from 1988. The evidence appears to support variants of hypotheses (1) and (2), but not hypothesis (3). It is proposed that serious interest or involvement in astrology is not primarily the result of a lack of scientific knowledge or understanding; rather, it is a compensatory activity with considerable attractions to segments of the population whose social world is labile or transitional; belief in astrology may be an indicator of the disintegration of community and its concomitant uncertainties and anxieties. Paradoxical as it may appear, astrology may be part and parcel of late modernity.

1. Introduction

Across the industrialized world, astrology has attractions for large numbers of people. Horoscopes are read by millions; astrologers are personally

---

[1] London School of Economics, Department of Social Psychology

[2] Science Museum and Imperial College, London

consulted by tens or hundreds of thousands; rumour has it that the London City is a booming place for astrological consultancy; even the wives of Presidents[1], it appears, may consult with astrologers before advising their husbands on how to conduct affairs of state. In all these situations astrology seems to offer a degree of certainty where uncertainty prevails. To many scientists and science educators who are concerned about the public understanding of science, the enduring popularity of astrology is an affront. How can it be, they ask, that in the last decade of the 20th century so many people are still prepared to embrace pre-scientific and even frankly superstitious belief systems?

Faced with the task of accounting for the enduring popularity of astrology, it is tempting to invoke the phenomenon of 'anti-science' - that is, active resistance to the principles and practises of science. In this context, it may be significant that the first of a series of US-Soviet conferences on the social and political dimensions of science and technology, which was held at the Massachusetts Institute of Technology in May 1991, was devoted to 'Anti-Science Trends in the United States and the Soviet Union'. Significantly, the two parallel keynote addresses to this conference - by Gerald Holton, of Harvard University, and Sergei Kapitza, of the Institute for Physical Problems (Moscow) - pointed to the need for a critical understanding of the phenomenon of anti-science. According to Holton anti-science in the US is symptomatic of a long-standing struggle over the legitimacy of the authority of conventional science;[2] while for Kapitza, anti-science in the east is part and parcel of the wider social and political transformation of the former Soviet Union.[3]

In a recent BBC radio programme prominent representatives of churches, science, and the arts discussed the apparent popularity of astrology and parasciences in Britain under the label 'pre-millennium tension' [PMT].[4] Ironically, on the issue of astrology and parasciences, the traditionally polarised positions of science and religion converged. It seems that present day astrology claims the territory which makes the Church and Science equally nervous. Albeit, the nervousness may have different sources.

In this paper, we investigate the phenomenon of popular belief in astrology in Britain in the late 20th century. Our evidence concerning the place of astrology in British culture is derived from the results of a 1988 national random sample survey designed to estimate levels of public interest in, understanding of and attitudes towards science and technology. In the course of this survey several questions were asked about astrology[5].

The results of these items enable us to explore three different sociological hypotheses which have been advanced to account for the prevalence of belief in astrology amongst certain sections of the public: first, that astrology is attractive to people with intermediate levels of scientific knowledge [superficial knowledge hypothesis]; second, that astrology is attractive to people who possess what has been termed 'metaphysical unrest' without integration into a Church; their unrest could therefore be considered free-floating [metaphysical unrest hypothesis]; and third, that belief in astrology is prevalent amongst people with authoritarian personality characteristics [authoritarian personality hypothesis].

Astrology must be considered the "grandmother" of modern science in at least two aspects: its concern with regularities in the universe, and its attempt to deal with these regularities numerically. Keith Thomas observed that 'at the beginning of the 16th century astrological doctrines were part of the educated man's picture of the universe and its workings'; London was a booming centre of astrological divinations for a mainly elite clientele of Court, nobility and Church until its decline in the mid-17th century.[6] In one sense it is not surprising that in a country that prides itself on tradition and continuity we find residuals or even revivals of such activities in the late 20th century. In this paper we try to locate contemporary belief in astrology in order to understand its social and psychological functions; while temporarily abstaining from evaluations of the belief itself.

We begin by defining our measures of public belief in astrology, and then proceed to use these measures to explore the three hypotheses.

2. Measuring Popular Belief in Astrology

The British survey was conducted in the early summer of 1988. The sample of 2009 respondents was designed to be representative of the adult population of Britain over the age of eighteen. The survey was conducted by means of face-to-face interviews lasting between forty minutes and one hour. The questionnaire covered a wide variety of topics in the general field of science and technology. In particular, it developed a multi-item scalar measure of scientific understanding. Further details of the survey methodology and the results on public understanding of science have been published elsewhere.[7,8,9,10]

So far as the present study is concerned, the following items from this national survey are of particular interest. First, respondents were asked 'Do you sometimes read a horoscope or a personal astrology report?'. Those who responded positively were then asked (a) how often they read a horoscope or personal astrology report [frequency] and (b) how seriously they took what these reports said [seriousness]. 73% of respondents claimed to read a horoscope or personal astrology report. 21% said that they would read it 'often', 23% 'fairly often', 29% 'not often', and 27% did not read it 'at all'. Hence, 44% claimed to do so often or fairly often. However, a rather smaller number of respondents (6%) claimed to take what horoscopes or personal astrology reports said either 'seriously' or 'fairly seriously'. 67% took it not very seriously, and 27% took it not at all seriously. This result points immediately to the problematic status of astrology in the minds of many of those who take at least some personal interest in it.

Figure 1: the combined percentages of respondents for two questions: 'how frequently do you read astrology columns?' and 'how seriously do you take it?'

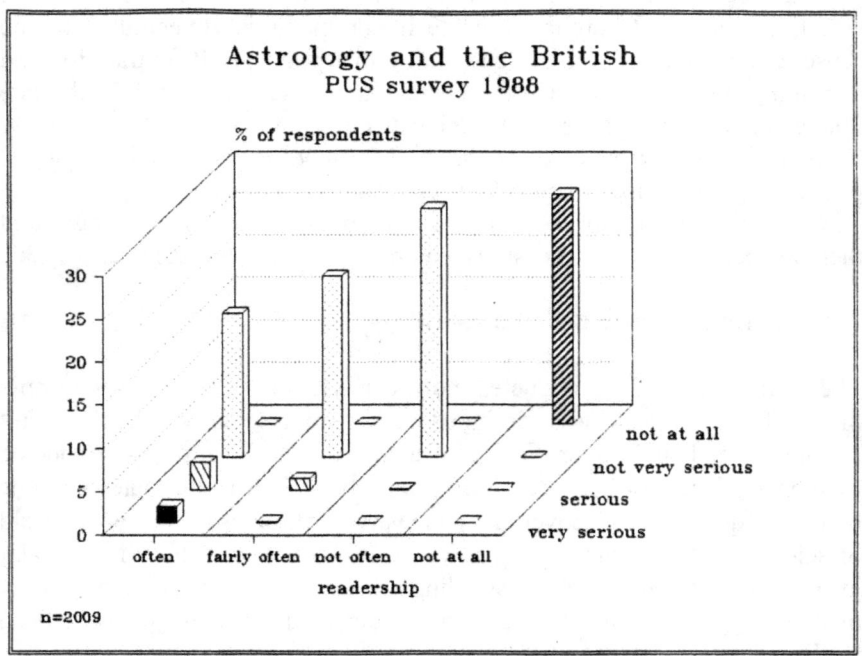

In order to accommodate these results in a useful way, we have combined them into a single scalar measure. Figure 1 brings together the results on readership and seriousness, which we combined into a 5 point-scale of belief in astrology. The scale is derived from the readership and seriousness results in the following way: those who reported that they read horoscopes often or fairly often and that they took them seriously or fairly seriously are ranked 5 (serious believers 5%); those who reported that they read horoscopes often and that they took them not very seriously are ranked 4 (non-serious believers, 18%); those who reported that they read horoscopes fairly often and that they took them not very seriously are ranked 3 (non-serious believers, 21%); those who reported that they read horoscopes not very often and that they took them not very seriously are ranked 2 (non-serious believers, 29%); and those who reported that they did not read horoscopes at all are ranked 1 (non-believers, 27%).[11] With around 5% of the population or 2.5-3 million, the constituency of serious believers in astrology is a small minority compared to the constituency adhering to basic religious creeds such as 'God', a 'life after death' or 'miracles', which includes half or more of the British population.[12] For much of the following analysis the 5-point scale is reduced by pooling 1+2, 3, and 4+5 into a 3-point scale.

Another item in the survey invited respondents to estimate the scientific status of astrology (which was defined as 'the study of horoscopes') on a 5 point-scale, from 'not at all scientific' to 'very scientific'. 32% of respondents stated that astrology was not at all scientific (scale point 1), while 13% stated that it was very scientific (scale point 5); 18% said it was in between (scale point 3); a further 17% tended towards 'not scientific' (scale point 2), and 14% tended towards 'scientific' (scale point 4); 5% did not know.

Our survey incorporated two standard measures concerning religious belief and religious integration. Religious belief was constructed as a scalar measure on the basis of responses to the following agree/disagree items: 'spiritual values taught by religion are important'; 'there is no such thing as a God'; 'people should rely more on the power of prayer'; and 'Adam and Eve never really existed'.[13] Religious affiliation was constructed as a scalar measure on the basis on the following items: 'Do you regard yourself as belonging to any particular religion?'; and (if yes), 'Apart from such special occasions as weddings, funerals and baptisms, how often nowadays do you attend services or meetings connected with your religion?'.[14]

Finally, the survey comprised two standard scales on 'authoritarianism-egalitarianism' and 'social efficacy'. Authoritarianism is indicated by consistently agreeing with statements such as 'censorship of film and magazines is necessary to uphold morality' or 'school should always teach children to obey authority'. Social efficacy is indicated by disagreeing with statements such as 'I feel it's very difficult to have any real influence on what other people do or think' or agreeing with 'people like me can influence the government by taking an active part in politics'.

3. Exploring the Basis of Popular Belief in Astrology

Equipped with the measures that have been described above, we can begin to explore the basis of popular belief in astrology. We shall do this by considering in turn three different hypotheses that have been advanced to account for this phenomenon.

i.   Superficial Knowledge

It has been claimed that belief in astrology is the product of a relatively slight or superficial acquaintance with the stock of modern scientific knowledge. On this view, people with what might be termed an intermediate level of scientific understanding may be attracted by astrology because it possesses many of the 'trappings' of orthodox science (systematic structure, predictive power, numeracy etc.); but they may be insufficiently well equipped to see that these things really are the 'trappings' rather than the substance of genuine science. Thus, in his classic paper of 1957 on the *Los Angeles Times* Astrology Column as an example of 'secondary superstition', Theodor Adorno wrote as follows:

> While the naive persons who take more or less for granted what happens hardly ask the questions astrology pretends to answer and while really educated and intellectually fully developed persons would look through the fallacy of astrology, it is an ideal stimulus for those who have started to reflect, who are dissatisfied with the veneer of mere existence and who are looking for a 'key', but who are at the same time incapable of the sustained intellectual effort required by theoretical insight and also lack the critical training without which it would be utterly futile to attempt to understand what is happening.[15]

We may pass over what seem by today's standards the somewhat elitist and patronising tones of Adorno's analysis. What concerns us here is whether the basic prediction - that astrology is attractive to people with intermediate levels of scientific understanding - is fulfilled. If that were the case, we would expect belief in astrology to be positively correlated with knowledge of science up to a certain level of scientific knowledge, beyond which this correlation becomes negative. In other words, we would expect a non-linear inverted U-shape relationship shown between scientific knowledge and the status of astrology.

This issue may be addressed by comparing the results of our question on the scientific status of astrology with the results of our multi-item scalar measure of scientific understanding. Figure 2 shows these results, compared with those for a similar item on the scientific status of physics. While there is a linear relationship between scientific understanding and the perceived scientific status of physics, there is a curvilinear relationship between scientific understanding and the perceived scientific status of astrology. In other words, our data do indeed bear out Adorno's hypothesis.

It should be noted that Figure 2 gives the proportions of respondents who ranked astrology and physics as 'very scientific'. We can learn a little more by comparing these results with those for other available options concerning the scientific status of astrology. Figure 3 shows the results for three groups of respondents: those who stated that astrology is not scientific (responses 1 + 2); those who stated that astrology is neither scientific nor unscientific, or who said they didn't know (neither + don't know); and those who stated that astrology is scientific (responses 4 + 5). Those with low levels of understanding have a strong tendency to avoid a definite judgement about astrology; while those with high levels of understanding have a strong tendency to state that astrology is unscientific. Amongst those with intermediate levels of understanding, there is less obvious consensus: some think astrology is scientific, some think it is not, and some don't know.

So much for the perceived scientific status of astrology. What, we may ask, about belief? Figure 4 compares belief in astrology with scientific understanding measured by a 28-item knowledge quiz.[16] As we might expect overall there is a negative correlation between scientific understanding and belief in astrology ($r = -.21$). However, on closer inspection it emerges that this negative correlation applies only to the

Figure 2: the scientific status attributed to physics and astrology in relation to the level of understanding of science; percentage of respondents saying 'scientific' or 'very scientific' combined.

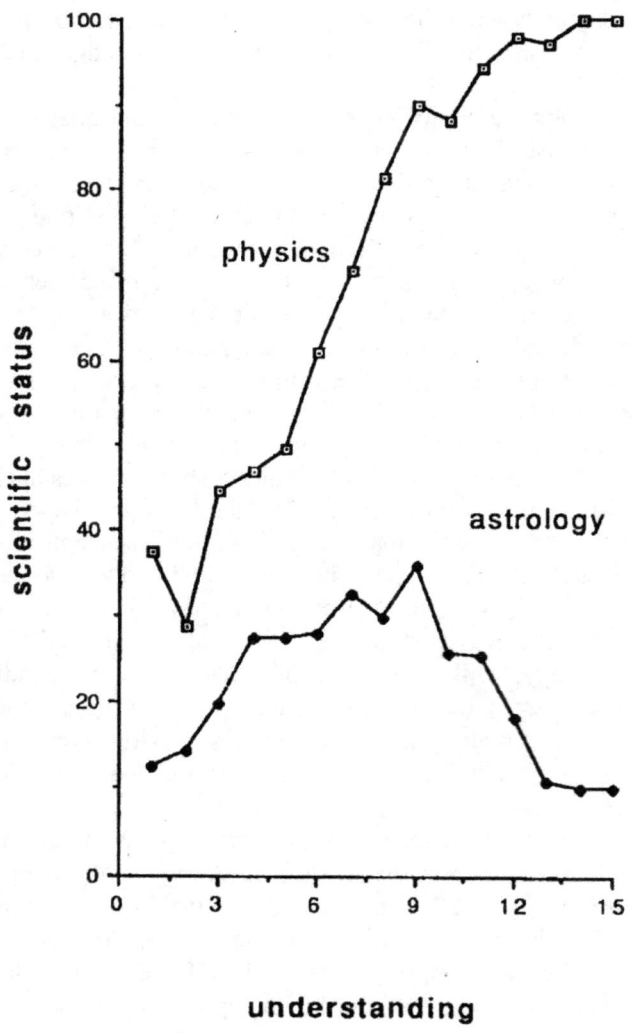

Figure 3 shows the percentage of respondents saying that 'astrology is not scientific', 'don't know' or 'astrology is scientific' in relation to levels of understanding of science.

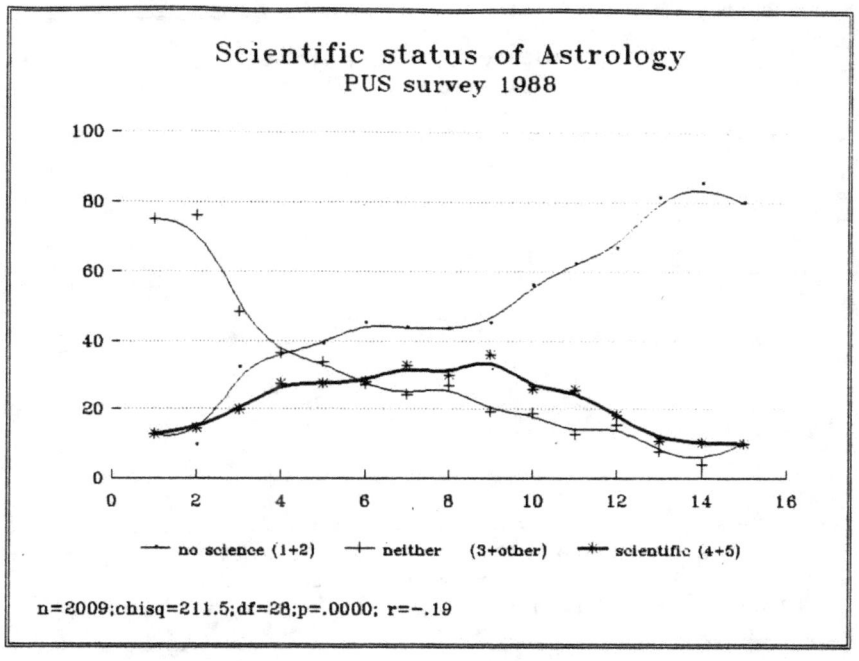

upper half of the understanding scale. We may wish to ignore the sudden jump of belief in astrology at the very top of the knowledge scale, which is based on a too few observations to be significant. However, within the 50% of the general public whose relative scientific understanding is below average, there is no correlation at all between levels of understanding and belief in astrology. This is a pointer to a potential problem with measures of scientific literacy that incorporate questions on the scientific status of astrology.[17] Empirically, astrology and science are not mutually incompatible at least at lower levels of scientific enculturation. To use astrology as a threshold measure for 'scientific literacy' may be justifiable on normative grounds, but it ignores the social phenomenon of compatibility or incompatibility between these two forms of knowledge, which is itself a significant cultural variable. We expect the correlation to differ across cultural contexts.[18]

Figure 4 shows the average intensity of belief in astrology in relation to the level of scientific understanding.

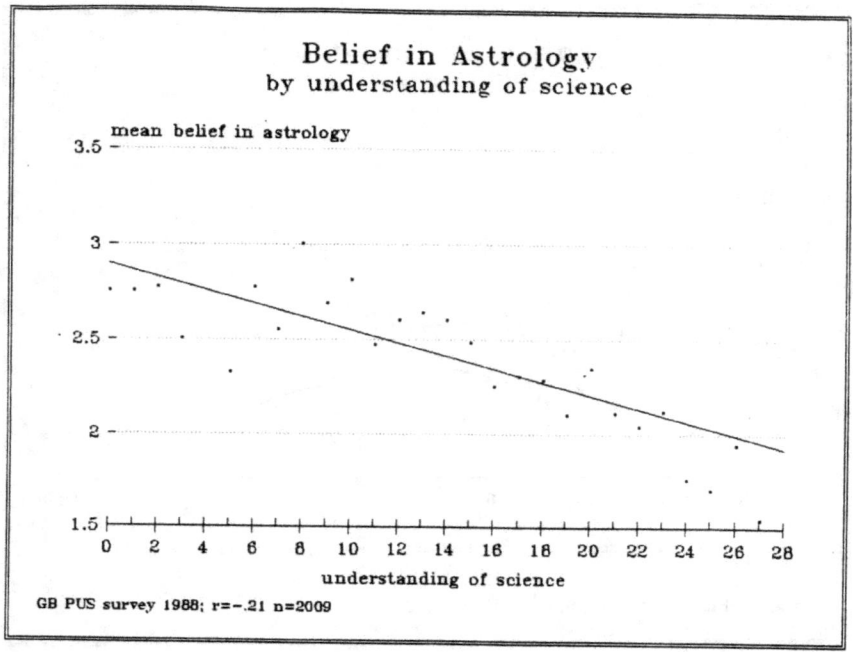

ii. 'Metaphysical Unrest'

It has been claimed that astrology has particular attractions for people who are alive to religion but who are poorly integrated into the institutional structures of a religious community. In this category are, for example, those who have been brought up in a particular religion and retain a religious outlook on life, but who for one reason or another (including social mobility or the collapse of community) have ceased to be closely tied to the particular church in which they were raised. Thus, Maitre and Boy & Michelat have observed in France of the 1960s and 1980s and Schmidtchen in Germany of the 1950s that astrology tends to be less popular amongst those who are closely integrated into the institutions of organized religion. The French characterize astrology as a petit-bourgeois phenomenon of social uncertainty, social isolation and individualisation.[19] According to Valadier, this result is consistent with the hypothesis that astrology feeds upon a free-floating 'metaphysical unrest', or a desire to recover a sense of the sacred and a sense of unity on the part of people whose life world no longer provides for these experiences; Pollack sees it as one among many forms of religiosity-outside-the-church in the context of the collapse of old certainties in Eastern Germany.[20] Based on these previous observations, we would expect to find serious inclinations towards astrology most prevalent among religious believers with little or no religious integration.

We may put this hypothesis to the test in the context of our British data. Our data show that there is a very slight tendency for belief in astrology to be greater amongst those with higher levels of religious belief ($r = 0.10$). However, inclination towards astrology is highest amongst those with intermediate levels of integration into the institutions of organized religion. Putting these results together, Figure 5 shows average belief in astrology in relation to both religious belief (1 = low; 3 = high) and religious integration (1 = low; 3 = high).

Figure 5: the average intensity of belief in astrology in relation to religious belief and religious integration

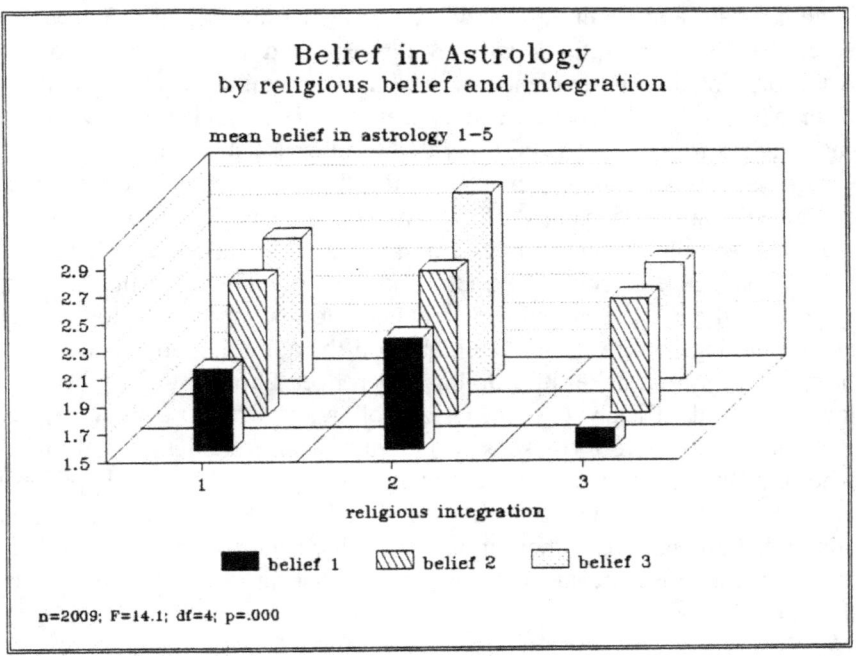

We see that belief in astrology is highest amongst those who combine strong religious belief and intermediate or low religious integration. The fact that an intermediary level of integration is associated with highest level of belief in astrology is perhaps unexpected. On the other hand, it may be that having one foot in the Church and the other outside it may be the very situation of social uncertainty which Valadier takes as diagnostic for present-day belief in astrology. To this extent, therefore, we are able to confirm Valadier's hypothesis and Schmidtchen, Maitre and Boy & Michelat's results suggesting that astrology has particular attractions for those who may be experiencing free-floating metaphysical unrest. Needless to say, our data do not permit us to explore the sources of such unrest in the lives of our respondents. This is an area where qualitative and biographical research may be more revealing.

iii. The Authoritarian Personality

The third and last hypothesis that we shall consider takes us back to the work of Theodor W Adorno. In the course of his analysis of astrology, Adorno noted that in general terms the astrological ideology resembles, in all its major characteristics, the mentality of the 'high scorers' of The Authoritarian Personality'. In addition to what he believed to be the narcissism, self-absorption, naive empiricism and fatalism of astrology, Adorno pointed to its tendency to attribute everything negative in life to external, mostly physical circumstances. In these and other ways, he suggested, astrology had affinities with the authoritarian personality.[21]

Once again, our data may be used to test this hypothesis since the survey contained a standard battery of psychological items designed to provide measures of authoritarianism-egalitarianism and 'social efficacy', defined as personal sense of control over the social world. The data shows that in our study there is no significant tendency for belief in astrology to be greater amongst those who score higher on the authoritarianism scale. We find, however, that belief in astrology is stronger amongst those who score low on social efficacy ($r = -.21$). Astrology, it would seem, is indeed particularly attractive to persons with certain characteristics, namely those who have little sense of control over their lives. Thus, Adorno's hypothesis is not supported by our data, while the fatalism element was confirmed. Given that this famous authoritarian personality syndrome is more complex than our crude measure suggests, we suggest that further work is needed on this subject.

4. Characterizing the Believers in Astrology

According to our results, the field in which the believers in astrology are generally to be found is one in which people possess intermediate levels of scientific understanding, high levels of religiosity, and low levels of religious integration. But what sorts of people are actually to be found within this field? In addition to what has already been said about personality, our data suggest that women are more likely to believe in astrology than men. Among the believers in astrology [scale 4+5] 77% are women; among the declared sceptics 73% are men [scale 1+2]. With the exception of clerks (a high proportion of whom are, of course, women) self-employed, skilled and semi-skilled workers are in that order more likely to believe in astrology than people in professional and managerial occupations. It is interesting to note here that according to Boy & Michelat, different social strata are associated with different sorts of 'para-interests': in France astrology is the pursuit of the less educated, while para-science is the pursuit of the highly educated. Our data do not allow us to compare this result with the situation in Britain.

These simple correlations are difficult to interpret because of the notorious problem of confounding variables. In other words, it may be that we find a correlation between belief in astrology and social class only because both in turn are related to some third factor (such as education, or social efficacy). To reduce the ambiguity of our results, we have subjected our data on belief in astrology to a form of statistical analysis (Logit modelling) which is designed to analyse differences between two unequally distributed groups.[22] In this case, we wish to analyse the contributions to differences in astrological belief of each of a series of independent variables. Each independent variable is assessed individually, whilst possible effects from all other variables are controlled. This analysis ranks independent variables in order of importance, and it excludes variables which are found to make no statistically significant contributions.

We used a Logit model in which differences between sub-sets of the sample with respect to belief in astrology were analysed with the following independent variables: interest in science; understanding of science; religious belief; religious integration; authoritarianism; social efficacy; age; gender; marital status; social class; educational level; and nature of work (i.e. full/part-time). Comparing the extreme groups of serious astrology believers (ranked 5) with non-believers (ranked 1 + 2) in this way, we obtain the following results. The variables which are relevant for the model are in order of importance: (1) gender, (2) religious belief, (3) living alone or in partnership, (4) age, (5) religious integration, and (6) the attributed scientific status of astrology. All other variables are irrelevant in explaining the difference between serious believers and sceptics. Note that the religious variables remain important, while personality and scientific understanding fall out of the equation. This indicates that the 'metaphysical unrest' hypothesis may be the strongest of the three hypotheses.

Comparing the category of playful, non-serious believers in astrology (ranked 4) with the sceptics (ranked 1 + 2), we obtain slightly different results. Again in order of importance the following variables are relevant: (1) gender; (2) marital status; (3) social efficacy; (4) educational level; and (5) attributed scientific status of astrology. In distinguishing between the playful and curious approach to astrology and the sceptics we lose the religious variable from the equation and gain education and efficacy.

At least as significant as the list of items that appear in these analyses is the list of items that do not. From these results, it would appear that interest in science and scientific understanding are not significant contributors to variations in belief in astrology. This, in turn, casts serious doubt on the advisability of employing measures of belief in astrology as constituent items in larger constructs concerned with scientific understanding or scientific literacy.

On the basis of these results, we can risk a caricature of believers in astrology. Serious believers in astrology tend to be: female rather than male; single rather than living with partners; younger rather than older; and religiously motivated rather than indifferent; and inclined to attribute scientific status to astrology. The non-serious and playful astrology consumer also tends to be female and to live alone, to be less educated, less in control of their affairs, and to consider astrology to be more scientific than the sceptics allow.

5. Conclusion

We began by citing recent concerns at the rise of astrology as an anti-science phenomenon, East and West. Kapitza suggests that in part the rise of anti-science in the (former) Soviet Union may be explicable in terms of the ideological collapse of the Soviet empire. Such a collapse may be expected to have left an intellectual and spiritual vacuum, and this in turn will have helped to bring about a certain amount of social disintegration. Similarly, Holton proposes that the anti-science phenomenon in the United

States should be understood as part of a deeper opposition both to the authority of science and to a certain conception of modernity. Both of these analyses invite us to consider popular belief in astrology as a great deal more than the passive result of mere ignorance.

In general, we suggest that there are three different ways of approaching the problem of popular belief in astrology. First, it may be regarded positivistically, as an anachronistic survival of a pre-scientific world-view. In this context, popular belief in astrology is seen as an atavistic phenomenon. Second, it may be regarded anthropologically, as an alternative world-view deserving of attention and respect in its own right. In this context, we are required to make no value-judgements about the respective merits of non-scientific and scientific positions. Third, it may be regarded sociologically, as one among a number of potential compensatory activity that may be attractive to individuals who are struggling to come to terms with the uncertainties of life in late modernity.

In this paper, we have inclined towards the last of these approaches. Belief in astrology is rather a matter of the moral fabric of modern society than of scientific literacy. It seems that in Britain, as in Germany or France, belief in astrology is prevalent among particular social groups; groups which, as we have indicated, may be experiencing difficulty in accommodating their religious feelings to life in an uncertain post-industrial culture. Paradoxical as it may seem, therefore, we conclude that popular belief in astrology may be part and parcel of late modernity itself.

References

1. An earlier version of this paper was given to the Annual Meeting of the American Association for the Advancement of Science in Chicago, 9 February 1992; at the time it was common currency that Nancy Reagan, the wife of former President Reagan, was consulting with astrologers on matters of US state affairs.

3. Kapitza S (1991) 'Anti-science trends in the USSR', *Scientific American*, 265, 2, August, 18-24.

4. 'Moral Maze', 14 November 1996, BBC4, 9.00-10.00; moderated by Melvin Bragg. This term is a slightly sexist pun on the medically controversial 'pre-menstrual tension'.

5. Acknowledgement: The 1988 British national survey of public understanding of science is a joint Science Museum/University of Oxford and Community Planning Research (SCPR) project funded by the Economic and Social Research Council, grant numbers: A 09250013 and A 418254007.

6. Thomas K (1971) *Religion and the Decline of Magic. Studies in popular Beliefs in the 16th and 17th Century*, London, Penguin, 337ff.

7. Durant J R, Evans G A and Thomas G P (1989) 'The public understanding of science', *Nature*, 340, 11-14.

8. Evans G A & J R Durant (1989) *Understanding of Science in Britain and the USA, British Social Attitudes: Special international report*, edited by R Jowell et al., Aldershot, Gower, 105-119.

9. Durant J R , G A Evans and Thomas G P (1992) 'Public Understanding of science in Britain: the role of medicine in the popular representation of science', *Public Understanding of Science*, 1, 2, 161-183.

10. Evans G and J Durant (1995) 'The relationship between knowledge and attitudes in the public understanding of science in Britain', *Public Understanding of Science*, 4, 1, 57-74.

11. To measure the internal consistency of the 'belief in astrology' scale we use Cronbach Alpha's Reliability Coefficient: 0.92. Alpha is a measure for the covariance among all the items in the scale; Cronbach L J (1951) Coefficient alpha and the internal consistency of tests, *Psychmetrica*, 16, 297-334.

12. Greeley A (1992) 'Religion in Britain', Ireland and the USA, in: R Jowell, L Brook, B Prior, B Taylor (eds) *British Social Attitudes*, the 9th report, Altershot, Dartmouth, 51-70.

13. Internal consistency of the religious belief measure: Cronbach Alpha's reliability coefficient = 0.70.

14. Internal consistency of religious integration: Cronbach Alpha reliability coefficient = 0.73.

15. Adorno T W (1957) 'The stars down to earth, the Los Angeles Times astrology column, a study in secondary superstition', *Jahrbuch fuer Amerikastudien*, Heidelberg, 2, 19-88, reprinted in R.Adorno, *The stars down to earthand other essays on the irrational in culture*, edited Stephen Crook (Routledge, London and New York, 1994).

16. See Durant, Evans, and Thomas (1989) op.cit.

17. Miller J D (1983) 'Scientific literacy: a conceptual and empirical review', *Daedalus*, 112, 3, 29-48; Miller J D (1991) 'The public understanding of science and technology in the United States', 1990, A report to the National Science Foundation, Dekalb, Illinois, February 1991.

18. We do recall from the Chicago meeting in 1992 that in the discussion an Indian theoretical physicist was quite irritated and outspoken about the tacit assumption in much of the discussion according to which science and astrology were incompatible. He made

reference to the Indian context where Brahmanic knowledge traditions seem to have no problem of compatibility between modern science and astrology.

19. Maitre J (1966) 'La consommation d'astrologie dans la societe contemporaine', *Diogenes*, 53, 92-109; Boy D & G Michelat (1986) 'Croyance aux parasciences: dimensions sociales et culturelles', *Revue Francaise de Sociologie*, 27, 2, 175-204; Schmidtchen G (1957) 'Soziologisches ueber die Astrologie', *Zeitschrift fuer Parapsychologie und Grenzgebiete der Psychologie*, 1, 47-72.

20. Valadier P (1987) *L'eglise en proces. Catholicism et societe modern*, Paris, Flammarion; Pollack D (1990) 'Vom Tischrucken zur Psychodynamik. Formen ausserkirchlicher Religiositaet in Deutschland', *Schweizerische Zeitschrift fuer Soziologie*, 1, 107-134.

21. Adorno T W , Frenkel-Brunswik E, Levinson D J and Sanford R N (1950*)* *The Authoritarian Personality*, New York, Harper.

22. Aldrich J H and F D Nelson (1989) *Linear probability, logit and probit models*, New Bury Park, Sage.

BOOK REVIEW

Visions of the Future: Almanacs, Time, and Cultural Change, **Maureen Perkins**, Oxford: Clarendon Press, 1996, 270 pp.; hardback; £40; ISBN: 0-19-812178-4

This is a brilliant book, combining thorough scholarship with original insight. It should deepen our understanding of a remarkable number of subjects. Essentially, it concerns a key part of the process of rationalisation that has been so instrumental in producing what we now recognize as modernity.

Perkins has much to tell us about astrological almanacs, to whose importance Keith Thomas first alerted us; in this capacity, she builds on and extends the excellent work of Bernard Capp. There is also fascinating material here on comic almanacs and Australian almanacs - the latter including an example of cultural influence by a colony (in the person of James Ross) on metropolitan discourse.

But more important is her use of almanacs to gain access to the world of popular belief, and the tensions in its relationships with elite opinion. Here the pioneer was Peter Burke's *Popular Culture in Early Modern Europe* and Perkins' book easily holds its own with subsequent scholarship by David Vincent and others. (I can't help feeling it a pity, however, that she passed over E. P. Thompson's apt refinement of 'popular' as 'plebeian'.)

At the heart of her account is the campaign by the Society for the Diffusion of Useful Knowledge in the third decade of the nineteenth century against the Stationers' Company, monopolist publisher of almanacs. Led by Lord Brougham and Charles Knight, the SDUK targeted such long-standing annual titles as *Poor Robin, Partridge's* and especially *Vox Stellarum*, popularly known as *Moore's*, with its mysterious hieroglyphic and astrological prophecy. In 1800, a minimum of one person in every seven in England bought an almanac - which was read, of course, by several more - and far and away the most popular was *Moore's*. In 1838, its best year, it sold over half a million copies, netting the Company of Stationers £6,414.

Significantly, the editorial voice of *Moore's* was unimpeachably Whig, comprising a set of convictions shared by the SDUK. But the latter had correctly identified the former as a major site and source of 'the superstitions of the vulgar' (in the characteristic terms of the

Athenaeum, in 1828) and, as such, resistance to its desire to advance the scientific management and rationalisation of society. The goal, as Perkins puts it, was nothing less than 'a transformation of consciousness, from one which was connected to a pre-Enlightenment world of correspondences and humours perpetuated by popular almanacs, to one in which empirical observation and rational enquiry were the standard....[and] in which the natural world could be placed without recourse to 'irrational' concepts.' (p. 58)

Of course, the SDUK's empirico-rationalism was far from neutral, proceeding by a series of conflations linking 'useful', 'rational', 'scientific' and 'real'. In other words, this was a hegemonic struggle to replace one particular social construal of reality with another. (I should add, however, that Perkins is no wild-eyed student of cultural studies, however; a more sober and thoroughly documented account would be hard to imagine.) In this context, the dividing line between rationality and 'superstition' was bitterly contested. In a fascinating chapter on weather, Perkins tells the sad story of Admiral Robert Fitzroy, who pioneered efforts to take its prediction out of the hands of countrymen, astrologers and amateurs. In 1865, harried mercilessly by the press as a covert weather-prophet (and by astrologers on his other flank), Fitzroy took his own life.

Predictably, the overall results of the SDUK campaign were uneven and complex. In 1872, *Moore's* finally dropped the astrology, only to be severely punished by readers: sales dropped steadily to only 50,000 in 1895. It was farmed out to another publisher in the early twentieth century who re-introduced 'the voice of the stars', and still appears annually, though with nothing like its former circulation or influence. Meanwhile, in the 1830s, judicial astrology re-appeared in the metropolitan heartland, courtesy the new almanacs of Zadkiel and Raphael. These had a middle-class readership, and Perkins underestimates the significance of their success, which astounded Charles Knight; she could have made more use of them in grasping the complexity of mid-nineteenth-century middle-class discourse. She also succumbs to the temptation (which seems nigh-well irresistible to historians in this field) to perceive the 'death of popular astrology', this time in 1869-70 (p. 119); the evidence to the contrary in every daily tabloid newspaper, and even some broadsheets (to the disgust of others). True, Sun-sign columns aren't precisely early modern moon- and star-lore; but they are much closer to it than to the highly individual analyses of judicial astrology.

Overall, however, Perkins is right to hand the palm to the reformers. Their relative victory was apparent in the new breed of almanacs such as *Whittaker's*, advancing a concept of time - and this is central - that was algorhythmic, quantitative and clock-based. Banished to the social and intellectual margins - where it still survives - was the old communal, qualitative time incorporating planetary and lunar cycles, and their corollaries in the annual seasons.

This raises the question of whether a 'post-modern' suspicion of science, ecological crisis in our relations with nature, and a post-Newtonian quantum physics signal the imminence of a new popular sense of time, one that may have significant continuities with premodern cycles and qualities. Whatever the outcome, future historians will have to consult Perkins before setting out.

Patrick Curry

NOTES ON CONTRIBUTORS

Robin Heath is an astronomer and the author of *A Key to Stonehenge*, Bluestone Press, 1993. He was formerly a senior lecturer in mathematics and engineering and is the founder of Megalithic Tours, Cwm Degwel, St. Dogmael's, Cardigan SA43 3JF, UK.
Norriss S. Hetherington is director of the Institute for the History of Astronomy at the University of California, Berkeley, and the editor of the *Encyclopedia of Cosmology* (Garland Publishing, 1993).
Alan S. Weber is CEMERS Associate Fellow, State University of New York, Binghamton, teaching in the English department. He is currently a Visiting Assistant Professor at Pennsylvania State University.
Ken Negus was for many years Professor of German at Princeton. He is the author of *Kepler's Astrology: Excerpts*, Princeton, 1987.
Martin Bauer is a lecturer in Social Psychology and Research Methodology at the London School of Economics. His research focuses on 'resistance to change' and on the 'public understanding of science'. His publications include *Resistance to New Technology* (ed.), (Cambridge University Press, 2nd ed. 1997).
John Durant is Assistant Director of the Science Museum and Professor of the Public Understanding of Science at Imperial College, London. He is the founder editor of the quarterly journal *Public Understanding of Science*. He is also Chairman of the European Federation of Biotechnology Task Group on Public Perceptions of Biotechnology and is a member of the UK Government's Advisory Committee on Genetic testing.
Patrick Curry is the author of *A Confusion of Prophets: Victorian and Edwardian Astrology* (Collins and Brown, 1992) and *Prophecy and Power: Astrology in Early Modern England* (Polity Press, 1989). He is the editor of *Astrology, Science and Society* (Boydell Press, 1987).

DIRECTORY OF USEFUL ADDRESSES

Journal for the History of Astronomy, Science History Publications Ltd., 16 Rutherford Road, Cambridge, CB2 2HH, England.

The Center for Archaeoastronomy, PO Box 'X', College Park, MD 20741-3022, USA. tel: (301) 864-6637, FAX (301) 699-5337. The Center's newsletter also carries news of the International Society for Archaeoastronomy and Astronomy in Culture. <http://www.wam.umd.edu/~tlaloc/archastro/>

Traditional Cosmology Society, Dr. Emily Lyle, School of Scottish Studies, 27 George Square, Edinburgh, EH8 9LD, UK.

British Astronomical Association, Historical Section, Anthony Kinder, 16 Atkinson House, Catesby Street, London SE17 1QU.

Ascella Books, 3 Avondale Bungalows, Sherwood Hall Road, Mansfield, Nottinghamshire, NG18 2QJ, England (reprints of old astrological texts).

Pratum Book Company, PO Box 985, Healdsburg, California 95488, USA, tel (707) 431-2634, Fax (707) 431-0575, E Mail knowledge@pratum.com (extensive range of rare and out of print books on mystical cosmology).

Events:

Astrological Lodge History Seminar, 1 November 1997, 10.00 am-5.30 pm, 50 Gloucester Place, London W1, including Nick Campion on Rudolf Hess' use of astrology, Silke Ackermann on 'Astrology and Scientific Instruments', Judith Kolbas on 'Solar Supremacy or Royal Iconography', Annabella Kitson on 'Lilly's "Mock Suns" and "World's Catastrophe"', Caroline Gerard on the Lauriston Horoscope, Tomas Gazis on Byzantine Astrology. Tickets £10 members, £12 non-members, from Astrological Lodge of London, 50 Gloucester Place, London W1 3HJ.

CULTURE AND COSMOS

Culture and Cosmos is published twice a year, in spring/summer and autumn/winter.

Contributions and editorial correspondence should be addressed to Nicholas Campion, The Editor of *Culture and Cosmos*, PO Box 1071, Bristol BS99 1HE, UK, E Mail <culture@caol.demon.co.uk>.

Deputy Editor: Patrick Curry, Ph.D.

Editorial Board:
Dr. Silke Ackermann, Professor Anthony F. Aveni, Dr. Guiseppe Bezza, Professor. J. Bruce Brackenbridge, Dr. David Brown, Dr. Charles Burnett, Dr. Hilary M. Carey, Dr. John Carlson, Professor Robert Ellwood, Dr. Germana Ernst, Dr. Ann Geneva, Dr. Jacques Halbronn, Robert Hand, Professor Norris Hetherington, Professor Michael Hunter, Professor Ronald Hutton, Annabella Kitson MA, Dr. Nick Kollerstrom, Dr. Edwin C. Krupp, Dr. J. Lee Lehman, Professor Kenneth Negus, Professor John North, Professor P. M. Rattansi, Professor Francesca Rochberg, Professor James Santucci, Robert Schmidt, Professor Richard Tarnas, Dr. David Ulansey, Robin Waterfield, Dr. Charles Webster, Dr. Graziella Federici Vescovini, Dr. Paula Zambelli, Robert Zoller.

Copy Editor: Ian Tonothy
Technical Assistance: Sean Lovatt

Subscriptions:
Individuals £13 UK and Europe, £15 overseas*
Institutions £22 UK and Europe, £24 overseas
Payment should be by sterling cheque or money order, or Eurocheque, visa or mastercard. For credit card payments we need the full card number, address and name on card, and expiry date.
Subscriptions can be received by E Mail on <subs@caol.demon.co.uk>

*Members of the British Astronomical Association, The Astrological Association and The Historical Association are entitled to a discount. Please enquire.

Contributors Guidelines: Please see inside back cover.
Copyright of signed articles and correspondence remains with the authors.
Copying: Apart from fair dealing for the purposes of research or private study, or criticism or review, as permitted under the Copyright, Designs and Patents Act 1988, no part of this publication may be reproduced, stored or transmitted in any form or by means without the prior permission of the Publisher.

The cover shows the explorer Gonsales travelling to the Moon, from Francis Godwin's *The Man in the Moone* (1618).

Published by Culture and Cosmos, PO Box 1071, Bristol BS99 1HE, UK.
© Culture and Cosmos 1997
Printed by Cromwell Press Ltd., Broughton Gifford, Melksham, Wiltshire SN12 8PH

Editorial

Astronomy is more than the science of the stars. It is intimately connected to our ideas of our selves, our purpose and place in the universe. Currently it is fueling myths, beliefs and ideologies as much as at any time in its history. We have the neo-paganism based on such sites as Stonehenge and Avebury. There is the belief in UFOs, borne out of modern science and science fiction, which in the last decade has exploded in ever more detailed doctrines of alien visitation and abduction. On a more sober level we have the manipulation of the space race to provide propaganda for the cold war. In the post 1989 world the USA continues to use its space programme to reinforce images of the American dream; space has become the new frontier, and 1997's Mars Rover mission was designed to touch down on 4 July, the anniversary of the declaration of independence. Then there is the social impact of the new science, explored in terms of quantum physics by Danah Zohar and Ian Marshall.[2] If the political impact of Copernicus and Newton is now well established, then we have hardly begun to question what effect Einstein may have had on twentieth century politics.

Then there is the question of belief in astrology which, in the modern world, is perhaps as strong as it ever has been. Its popularity strikes many people as something of a historical problem: how in a modern rational, scientific world could such an irrational, unscientific belief flourish? Given that such words as rational and scientific are highly problematic and open to differing interpretations, astrology's contemporary revival does require study and explanation.

The significance of astrology to the history of ideas, religion and science is no longer in question. Lynn Thorndike's epic *History of Magic and Experimental Science* has made the argument convincingly. A select group of other scholars have examined its role in specific periods and cultures, notably in the ancient near east, the classical world and in medieval and Renaissance Europe. Yet while papers on the subject find their place in specialist academic journals, it is felt that the time has come for a journal which focuses specifically on its past and development, as well as on its social, political and religious functions.

Astrology can be defined as the use of celestial phenomena to interpret and predict events on earth. However, in order to flourish it requires a philosophical context based on a general belief that movements and changes in the heavens are significant for humanity, without the specific rules, technicalities and procedures necessary for astrological interpretation. This may be defined as cultural astronomy: the use of astronomical knowledge, beliefs or theories to inspire, inform or influence social forms and ideologies, or any aspect of human behaviour. Cultural astronomy also includes the modern disciplines of ethnoastronomy and archaeoastronomy.

The problem of definitions was dealt with by Michael Hoskin in his 1996 review of *Astronomies and Cultures*.[1] He posed the question "What astronomy is *not* an astronomy in a culture?" This is a valid point, and one may take a narrow or broad definition of the term cultural astronomy. In part the solution is one of emphasis. *Culture and Cosmos* will emphasise the cultural aspects of astronomy rather than the strict history of mathematical and technical astronomy, areas ably catered for elsewhere.

Culture and Cosmos will cover a broad spectrum of ideas: any and all of the ways in which human beings manipulate, exploit, analyse and interpret the heavens in order to understand, regulate and predict their individual concerns and collective lives. We may not be able to answer all questions, but at least we can ask them.

References

1. Hoskin, Michael, review of *Astronomies and Cultures*, ed. Clive L. N. Ruggles and Nicholas J. Saunders (University of Colorado Press, Niwot, Col., 1993), *Archaeoastronomy*, number 21, supplement to the *Journal for the History of Astronomy*, vol. 27, 1996, p 885-7.

2. Danah Zohar and Ian Marshall, *The Quantum Society*, (Bloomsbury Publishing, London 1993).

An Astronomical Basis for the Myth of the Solar Hero

Robin Heath*

Introduction
Our increasing knowledge of the megalithic culture of the British Isles in the 2nd and 3rd millennia BCE tends to confirm the proposition that megalithic astronomers measured celestial positions with considerable accuracy. The evidence indicates that they understood the 18.6 year nodal period and the moon's nine minute declination wobble.[1] They also had sufficient geometrical ability to re-proportion spacings between lines, divide circles into whole number polygons and divide lines into equal integer spacings.[2] We should therefore ask whether there is evidence of such early astronomy in the numbers which recur in certain myths. The following should be viewed as preliminary arguments.

The Thirty-Three Year Cycle
If our interest is megalithic astronomy then we should search for relevant evidence in the myths of the British Isles. One of the most recurrent numbers in the stories of the Tuatha de Danaan who, according to tradition, inhabited Ireland before 1,500 BCE, is thirty-three.[3] We are told, for example, that the first battle of Mag Tuired was fought by the saviour-hero Lug and thirty-two other leaders. In the same vein, Nemed, another hero, reached Ireland with only one ship, while thirty-three were lost on the way; Cuchulainn slays thirty-three of the Labriads in the Bru battle whilst a late account of the second battle of Mag Tuired names thirty-three leaders of the Fomorii, thirty-two plus their highest king.[4]

This material contains one clear and obvious common theme. Repeatedly, it reinforces an originally oral message which told the knowing listener to look to the number thirty-three as something relevant to a hero, a saviour. In his analysis of the Welsh *White Book of Rhydderch*, N.L.Thomas writes that 'Both three and eleven were equally symbolic, the multiplicand thirty-three particularly so. It has frequently been used to imply supra-human attributes, regal authority and deification.'[5]

---

* Megalithic Tours, Cwm Degwel, St.Dogmael's, Cardigan, SA43 3JF, UK.

We find evidence of the astronomical and mythical significance of the thirty three year cycle in other cultures. Perhaps the best known example is found in the tradition that Christ began his ministry at age thirty and was crucified at age thirty three.[6] The solar tradition in early Christianity is well-recorded, with the widespread identification of Christ with Helios and the fixing of Christmas to coincide with the festival of Sol Invictus, a few days after the winter solstice, and Easter close to the spring equinox.[7] It is partly on the basis of such evidence, together with the argument that epic myths such as those of Gilgamesh and Hercules represent solar cycles, that modern comparative mythology has produced the notion of the solar hero.[8] I am arguing that the numerical evidence in the Celtic tales provides astronomical evidence that they too could be considered solar heroes.

Megalithic Astronomy and the Solar Cycle
Many megalithic standing stones have been shown to relate to extreme Sun and Moon rising and setting azimuths against the local horizon. While the link at some sites is tenuous it is beyond doubt in others, as for example at Stonehenge.[9] The practical solar year is 365 days long. I say *practical* because most reference books claim that there are 365 and a quarter days in the year, a confusion with the Earth's orbital period around the Sun. Every fourth year an extra day slips in to make it 366 days. In four years there are thus 1461 days. It is fairly easy to observe the Sun's behaviour and thereby measure this number.

An equinoctial Sunrise marker, of which many still exist in the British uplands will, each year, deliver the vernal equinox sunrise from a slightly different position on the horizon. The 'quarter day effect' means that the Sun, each year, is displaced about a quarter of a degree from the marker stone. During three years of observation, the Sun appears to be slipping every more away from the alignment until, at the fourth year, two remarkable and very observable things happen simultaneously: the Sun rises once more very close to the marker stone, whence the day count - the tally - for the year is found to be 366 and not 365 days. Observation does not stop there, and a good human eye can detect much more minuscule angular changes than a quarter of a degree from watching sunrises.[10] The truth about solar year measurements carried out at the equinox is that the result is always 365 days unless sustained observations are conducted over many years. In that case the result is 365.25 days, a figure which, under optimum conditions could be reached after four years (see Figure 1).[11] This figure is within eleven minutes of time of that for the tropical year (365.24219 mean solar days) and is almost identical to that for the civil, calendar, year (365.2425 mean solar days).

Figure 1

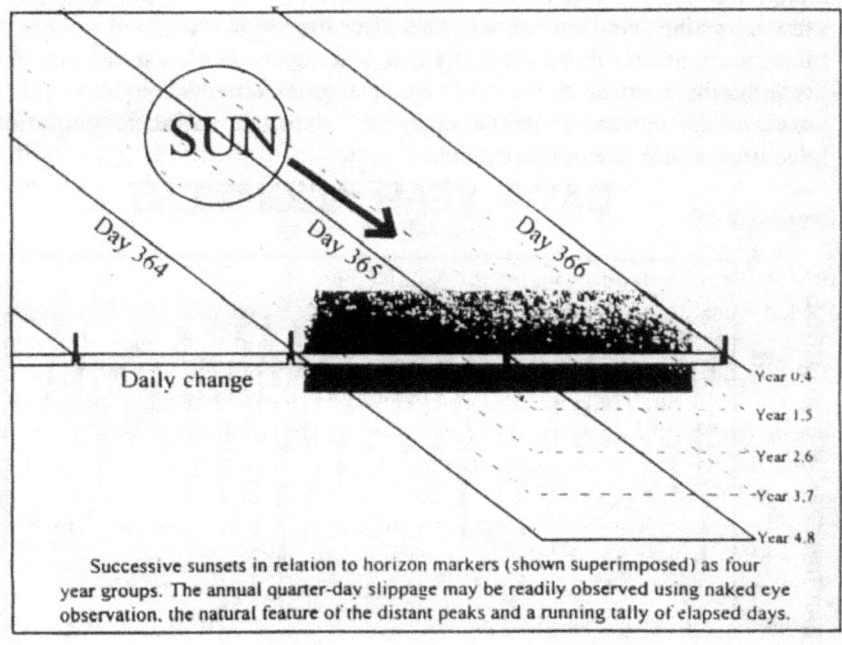

Successive sunsets in relation to horizon markers (shown superimposed) as four year groups. The annual quarter-day slippage may be readily observed using naked eye observation, the natural feature of the distant peaks and a running tally of elapsed days.

For longer time periods something else happens. Every once in a whole number of years the chance arises to measure the year with even greater precision. This can be achieved by observing certain key years when, once again, the Sun rises *precisely* behind the foresight - be this a stone marker or a distant mountain peak - in other words, *a perfect repeat solar cycle*.

What we may assume, courtesy of their enduring architecture, is that the megalithic astronomers could have readily evaluated the length of the solar year to two decimal places. They could accomplish this by marking 1461 equal lengths on a rope - the tally count of days in four years - and then folding it in half twice to obtain 365.25. The 1461 day tally is a

6 A Possible Astronomical Basis for the Myth of the Solar Hero

given, gleaned from simple observation and tally counting over four years.

After thirty-three years one can observe an <u>exact</u> repeat of the original equinoctial rising behind the marker stone (See Table 1, Figure 2). To a megalithic astronomer, this same phenomenon would have translated as an *exact repeat rising* (or setting) behind a marker, whilst a modern astronomer would note that the Sun's declination will be identical on the same calendar date thirty-three years after the value read from a book of tables today; thirty three years is a true solar cycle. If we can assume that the megalithic astronomers made exact angular observations over many years, as the current evidence suggests,[12] then this phenomenon would have been a familiar one to them.

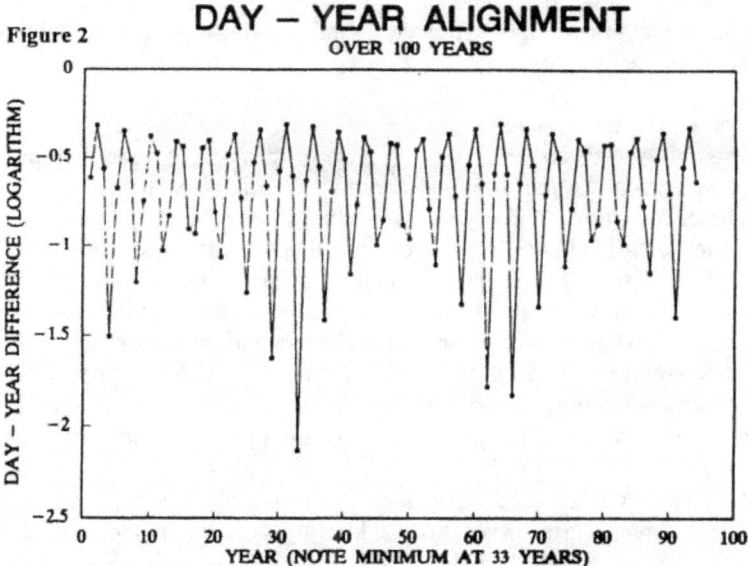

The vertical axis indicates the error from 364.242 days plotted on a logarithmic scale. The base line, 2.5, represents the minimum deviation. The horizontal axis indicates years from 0 to 100. This clearly indicates the thirty three year pattern. ***Diagram courtesy of Nick Kollerstrom.***

However one interprets the data, this presents a possible astronomical source to the use of the number thirty three in heroic myths. Thus when Christ's resurrection occurs at age thirty-three, witnessed as it was at sunrise, we may be faced with a sophisticated astronomical/calendar metaphor. Even the rolling away of the stone to reveal the resurrected saviour may plausibly be argued to represent the emergence of the Sun from behind a stone marker, inaugurating another thirty three year cycle.[13]

Conclusion

Contemporary archaeology has established the astronomical sophistication of the megalith builders to a previously unsuspected level. So far the arguments, naturally enough, have rested primarily on the archaeological record. However, a second line of argument may be derived from the mythical record, even though we have to account for the problem that the written sources are necessarily of a much later date than the stone remains. It is clear that myth may provide evidence of ancient astronomical knowledge, while astronomy may also provide an additional view of ancient myths.

Postscript: The Dresden Codex

Interestingly, we can find the solar cycle of thirty-three years within other cultures, such as the Mayan. I draw attention to this as additional evidence that thirty three-year cycles may have been apparent to early astronomers. The Dresden Codex, a collection of divinatory almanacs mostly tied to the 260 day divinatory cycle, contains eclipse timings compiled by the Maya and runs for almost 33 years.[14] This tabular codex abruptly finishes after 32 years and 270 days and the reason for its length is understood to be connected with the fact that 46 sacred periods of 260 days tally with this period.[15] However, it is also evident that after this time frame, which corresponds to 405 lunations, the Sun will meet the lunar node axis and produce an eclipse. In other words, 405 lunations is an eclipse cycle; after 32 years and 270 days, the nodes will have made one and three quarter revolutions whilst the Sun has made 32 and three quarter revolutions of the ecliptic, the result being that the Sun meets one of the nodes, resulting in an eclipse.[16] As the Dresden Codex is in part a document about eclipses, there was thus no real need to continue the tabulation to 33 years in order to complete the niceties of the evidently known solar cycle of thirty-three years, 405 lunations is the last eclipse possible within this cycle. It is therefore reasonable to infer that the codex recognises the importance of the 33 year solar cycle and that the Maya were familiar with it.

8 A Possible Astronomical Basis for the Myth of the Solar Hero

Table 1.

Important Solar Returns behind a horizon alignment			
Number of years	Days	Time Difference from whole number	Angular error from original solar observation
4	1,460.968	45 minutes	1 min. 30 sec.
21	7,670.086	124 minutes	3 min. 42 sec.
33	**12,052.992**	**10.7 minutes**	**0 min. 18 sec.**
62	22,645.016	23.53 minutes	0 min. 36 sec.

The Tropical Solar Year is 365.242199 days in length. Multiply this by whole numbers (of years) and look for products where the fractional part of the result tends towards zero or one. There are several contenders, shown above. Consecutive years contain an angular error of one quarter (15 minutes) of a degree. The Daily angular sunrise change along the horizon in Southern Britain at the equinox is over 0.7 degree. This is considerably more than one solar disc diameter (about 0.6 degree).

References

1. Alexander Thom, *Megalithic Sites in Britain*, (Oxford, Oxford University Press, 1967, pp 59, 165. Thom reckoned that good observation can detect angular changes as small as 2.5 minutes of arc (correspondence from Archie Thom to the author, 1994). John E.Wood, *Sun, Moon and Standing Stones*, (Oxford, Oxford University Press, 1980) wrote that 'the Temple Wood observatory shows inherent accuracies of declination measurement to around one hundredth of a degree'.
2. Wood, *Sun, Moon*, pp 36-56
3 The best source for the early myths is the Royal Irish Academy's edition of *The Book of Leinster*, 1880, a facsimile of the *Leabar na Nuachonghbhala* (also sometimes known as The book of Glendalough). This contains the earliest known version of the *Leabhar Gabhala*, the 'book of invasions', the primary source for the stories of the Tuatha de Danaan.. The most reliable version of this work is *Lebor Gabala Erenn: Book of the Taking of Ireland*, ed. R.A.S.MacAlister, Irish Text Society, 5 vols, 1938, 1939, 1940, 1941, 1954.
4. N.L.Thomas, *Irish Symbols of 3,500 BC* (Mercia Press, Cork, 1988), p 83.
5. Thomas, *Irish Symbols*, p 76. For The White Book of Rhydderch *see Llyfr Gwyn Rhydderch (The White Book of Rhydderch)*, with an introduction by T.M.Jones, (University of Wales Press, Cardiff, 1973, reprint of the 1907 edition edited by J.Gwenogvryn Evans) The original manuscript is National Library of Wales MSS Peniarth 4. Also see Rees, Alwyn and Brinley, *Celtic Heritage* (Thames and Hudson, London, 1961), pp 200 f., 318 ff., 338. Although *The White Book of Rhydderch* is dated to 1300-25, like other similar texts, it is widely believed to represent the written account of a much earlier oral tradition.

6. Luke 3.23, Acts 2, record that Christ's ministry began at age thirty.
7. The earliest reference to the celebration of Christmas on Sol Invictus, December 25, is the Philocalian Calendar of 336. There are different formulae establishing the celebration of Easter. The standard has become the Roman version: Easter Sunday is the first Sunday after the full moon following the spring equinox. See also Henry Chadwick, *The Early Church*, Harmondsworth, Middlesex, for discussion of Christian reverence for the sun in the 1st-3rd centuries.
8. The concept of the solar hero has been particularly popularized by the Jungians. Jung wrote that 'It is not enough for the primitive to see the sun rise and set; this external observation must at the same time be a psychic happening: the sun in its course must represent the fate of a god or hero who, in the last analysis, dwells nowhere except in the soul of man'; C.G.Jung, 'Archetypes of the Collective Unconscious', *Collected Works*, Vol. 9, part 1, p 6, trans F.R.C.Hull (Routledge and Kegan Paul, London, 1959).
9. For Stonehenge see Hugh Thurston, *Early Astronomy* (New York 1994), pp 45-55.
10. Thom, *Megalithic Sites*, p 108
11. Norman Lockyer argued that simple observation was sufficient to establish these figures: 'Had ignorance led to the establishment of a year of 360 days, yet experience would have led to its rejection in a few years...If observations of the Sun at solstice or equinox had been alone made use of, the true length of the year would have been determined in a few years'. Norman Lockyer, *The Dawn of Astronomy*, (Cambridge University Press, 1894) pp 245-6, reprint MSI, 1964.
12. At Loughcrew, in Ireland, Cairn F, Stone C1, there are a set of 62 inscribed markers, whilst nearby, at Fourknocks Passage, one may count three columns of eleven chevrons, totalling 33, picked onto a stone. See N.L.Thomas, *Irish Symbols*, p 73.
13. Matthew 28:1, 'as it began to dawn, towards the first day of the week'. In first century Greek astrology the first day of the week, and the first hour of the first day, were ruled by the Sun. If we can interpret the astronomical features of the resurrection story as an allegory of the solar cycle, then Mary, as the mother, represents the origin of the process, in other words the first measurement or alignment with the stone marker thirty three years previously. To extend the allegory, the stone blocking the tomb, the entrance to the underworld, rolls away revealing the resurrected form and his entrance back into the visible world.
14. See Floyd Lounsbury's paper in the *Dictionary of Scientific Biography*, Vol 15, supplement 1, Charles Gillespie general editor (Charles Scribener's Sons, New York)
15. Evan Hadingham, *Early Man and the Cosmos*, William Heinemann (London 1983), p 223. See also Anthony Aveni, *Skywatchers of Ancient Mexico*, (University of Texas, 1980).
16. See Thurston, *Early Astronomy*, p 201.

Early Greek Cosmology: A Historiographical Review

Norriss S. Hetherington*

Introduction

Early Greek cosmology has attracted much attention from classicists, historians, philosophers, and scientists, with each group bringing to the subject its own interests and biases. Purportedly authoritative reconstructions and analyses of ancient Greek cosmology exist in abundance, even though no philosophical writings of the Presocratic period, circa 600 to 400 BC, have survived. The Greeks' attempt to explain celestial phenomena in natural terms and to avoid supernatural or divine intervention is a common theme linking many otherwise disparate scholarly studies. A frequent point of dispute involves the degree to which ancient ideas are to be judged in the context of modern science.

From scientist-historians, by which I mean scientists who have become historians, we have, in the words of Victor Thoren, 'modern commentaries, written fairly uniformly over the last 150 years by men uniformly possessed of more astronomical ingenuity than historical perspective or critical sense. The result is a corpus of secondary material replete with literally incredible claims, many of them mutually (and some of them self-) contradictory.'[1] At the other end of the scholarly spectrum reside classicists, producing philological rather than philosophical or historical books. As William Stahlman wrote, 'The trees are here and accurately labelled, but we never see the forest.'[2]

Between these two extremes lie a few studies of ancient Greek cosmology in its cultural context, most often focusing on the Ionian school, or Milesians, chiefly Thales (c.600 BC), Anaximander (c.610-545 BC), Anaximenes (c.546 BC) and Heraclitus (c.540-c.480 BC), and Pythagoras (b.c.570/580 BC) and the Pythagoreans.[3] It is these who chiefly concern me here. However, neither most of the accounts of the Milesians nor the Pythagoreans adequately cover all the varieties of Presocratic cosmological thought, and historians are open to the charge of over-simplification, that 'for long enough we have thought of early Greek philosophy as a tennis match between Ionians and Italians, with all the Greeks in the middle gaping dumbly up as the ball flew to and fro above their heads'.[4]

Sources

* University of California, Berkeley,

One of the most useful and convenient collections of the raw materials for reconstructions and analyses is contained in *The Presocratic Philosophers* by Kirk, Raven, and Schofield. Their work includes extant fragments, mainly a few quotations from Presocratic works that have survived in books written after 400 BC, with the Greek original and English translation, and commentaries. Their principal sources are testimonia, comments in the writings, such as have survived, of Plato, Aristotle, and Theophrastus, written shortly after the Presocratic period; and the doxographical tradition, consisting of summaries of the works of Plato, Aristotle, and Theophrastus, and summaries of summaries, the primary source for the summarizers being the multi-volume history of early philosophy by Theophrastus.[5]

Yet none of the surviving sources is above question. The fragments probably are the most reliable, or least unreliable, but there is no surviving original against which to check them. The testimonia have an additional uncertainty. Aristotle gives serious attention to the earlier philosophers, but his judgment and his corresponding analysis and description of earlier philosophy may well have been distorted by his own belief in the importance of the material nature of the world. He desired 'to find predictions of his own conclusions in the works of his predecessors'.[6] Plato, in contrast, offers only casual remarks on his predecessors. Just how little has survived is illustrated by the example of Theophrastus. Of the approximately eighteen books he wrote, at most a handful have survived, and the doxographical tradition consists primarily of summaries of his books and summaries of summaries.

Classicists are virtually unanimous in doubting the reliability of the surviving sources, though with differing degrees of forcefulness. In *Early Greek Astronomy to Aristotle*, D. R. Dicks' argumentative scepticism left readers with a decidedly negative aftertaste.[7] He was even more polemical and irritating in journal articles and in 1966 he wrote that 'The literature is now full of references to the scientific achievements (so-called) of the Presocratics, and the earlier the figure (and consequently the less information of reliable authenticity we have of him) the more enthusiastically do scholars enlarge his scientific knowledge'...[8] Thales is the earliest such figure and Dicks considered that 'Inevitably there accumulated round the name of Thales, as that of Pythagoras (the two often being confused), a number of anecdotes of varying degrees of plausibility and of no historical worth whatsoever'.[9]

Speculation or Science?

Having dispensed with modern scholars, Dicks turned on their ancient subject, the Presocratics. He argued that 'Greek astronomy was still in the pre-scientific stage. Observations of astronomical phenomena...were rough-and-ready observations, unsystematically recorded and imperfectly understood, of practical men...whose main concern was to have some sort of guide for the regular business of everyday life...Ionian speculation seems to have taken very little note of such observation (some of its wilder flights of fancy might have been avoided, if it had taken more)...Not until Ionian speculation had played itself out and it was becoming increasingly obvious that such presumptive theorising bore little or no relation to the gradually accumulating stock of observational data, did mathematical astronomy even begin to develop'.[10]

In response to Dicks' characterisation of Ionian philosophy as a speculative enterprise without a scientific future and a philosophic sideline with no impact on the development of observational science or mathematical astronomy, one critic charged that what he offered was 'essentially a Baconian or neo-Baconian view of science which admits mathematical computation together with empirical observation as the necessary characteristics of science, but which denies any role to speculative hypotheses of a strongly theoretical nature'.[11] A classicist limiting himself to Greek and Roman subjects, Dicks did not look ahead to the influence of Pythagorean number mysticism on modern science and he gave short shrift to the cosmological fantasies of the Presocratics, rejecting sweeping statements from other scholars about supposed striking similarities between patterns of thought in ancient Greek and modern science. He did, though, concede that the Pythagoreans were beginning to move away from speculative thinking. Other scholars have seen in emerging Ionian rationalism the removal almost at a single stroke of the entire mythological scaffolding of earlier, pre-scientific thought.[12]

Ionian Rationalism
Focusing on the method rather than substance of Presocratic thought avoids the difficulty that most Presocratic theories are known to be false. To put it bluntly, as Jonathan Barnes did, 'none of the Milesian theories is true: the Milesians do not compose a Greek Royal Society; and their Transactions would not make any contribution to the sum of scientific knowledge'. Further, by focusing on the rational, philosophical element within Presocratic methodology rather than the mathematical, quantifiable element, historians can avoid the difficulty that 'none of the Milesians aspired to the sort of precision we require in a scientific theory:

their views are incurably vague; and underlying this vagueness is a complete innocence of the delights of measurement and quantification'.[13]

An emphasis on rationalism as a key characteristic of Presocratic cosmology also fits nicely with the 'Greek miracle' view of ancient history. Simply stated this holds that in the beginning there were 'charming but childish Egyptians and Sumerians with their weird and fantastic notions about the cow-goddess in the sky, the sweet waters under the earth, and so on, and then along came the Greeks who were adult rational people like ourselves'.[14]

Francis Cornford, a historian of ancient philosophy, did much to establish the belief in the miracle of Presocratic rationalism. In *From Religion to Philosophy*, first published in 1912 but still popular, he commented on the absence from Milesian philosophy of astrological superstition, magical powers, and mythical cosmogony. Less noted by readers is his conclusion that the advent of the new rational spirit was not a sudden and complete breach with the old, and that there remained a thread of continuity from science back to the supernatural world of the gods.[15]

Cornford's thesis has not been superseded, even if part of it, no doubt compatible with prevailing theories of race in England early in the twentieth century, now risks being found less than perfectly politically correct. 'The scientific tendency is Ionian in origin: it takes its rise among that race which had shaped Homeric theology, and it is the characteristic product of the same racial temperament.'[16] Cornford's Presocratics also aspired to be modern scientists. 'The aim of science...triumphantly achieved...succeeded in reducing physics to a perfectly clear, conceptual model, such as science desires...'[17] But the Presocratics could not have known what lay in the future. 'They were not trying to give a scientific system, since no one yet had told them what "science" ought to be.'[18]

By the 1930s Cornford had forged general agreement on the proposition that with Thales what we call western science first appeared in the world. He wrote that 'The intelligence became disinterested and now felt free to voyage on seas of thought strange to minds bent on immediate problems of action. Reason sought and found truth that was universal, but might, or might not, be useful for the exigencies of life...Science begins when it is understood that the universe is a natural whole, with unchanging ways of its own - ways that may be ascertainable by human reason, but are beyond the control of human action'.[19] With the rise of science, he thought, there occurred a corresponding demise of magic and mythology, themselves pre-scientific practices designed to bring supernatural forces under some measure of control.

Later, Cornford had second thoughts. All his life he was a dissenter, and ultimately he dissented against the proposition he had done so much to establish, that Ionian natural philosophers were scientific. Now he noted that their dogmatic pronouncements easily could have been upset by careful observation or the simplest experiment, but they had no empirical theory of knowledge to govern their speculations.[20] Published only posthumously, Cornford's second thoughts still have not entirely overtaken his first. A new generation of historians is more sensitive to the unscientific nature of Presocratic science, but some writers still single out rationalism, seemingly unaware both that they are echoing Cornford and that he has retracted much of the foundation upon which their derivative accounts are constructed. It is not inconsistent, however, to maintain that the Presocratics had a scientific world-view even though they lacked the experimental method.[21]

Benjamin Farrington was prominent amongst those who emphasised the rational in Presocractic cosmology. In his *Science in Antiquity*, he traced the development of ancient science in close relation to the history of philosophy. For the Milesians, he argued that Thales' importance lay in his being the first person known to have offered a general explanation of nature without invoking the aid of any outside power. He concluded that Anaximander's brilliant advance was toward a more abstract conception of nature; no longer was the underlying substance of the material world a visible, tangible state of matter, such as water, but the lowest common denominator of all sensible things arrived at by a process of abstraction.

Farrington placed the Pythagoreans, on the other hand, in a context of a spiritual revival brought about by the menace of the Persian advance, arguing that Pythagorean mathematics was primarily a religious exercise.[22] Here he was followed by other scholars, such as June Goodfield and Stephen Toulmin, who argued that 'Pythagoras, it is clear, was not so much the leader of a scientific research team, or the principal of an educational establishment, as (in modern terms) the guru of an Indian ashram'.[23] Earlier Bertrand Russell had found modern parallels for Pythagoras when he described him as 'a combination of Einstein and Mrs. Eddy'.[24]

Science and Society

Farrington forged a very different thesis in *Greek Science*, exploring connections with practical life, with techniques, and with the economic basis and productivity of Greek society. A Marxist, he was interested in the effect of class interests in determining early Greek philosophical opinion. Egyptian and Babylonian cosmogonies, known to Thales, had embodied the idea of water in the beginning, probably because the land in both countries had been won in a desperate struggle with nature by draining swamps. Thales, leaving out only the god who had let dry land be, still formed everything out of water. Farrington seemingly chose to ignore Aristotle's speculation that Thales may have arrived at his supposition from seeing the nurture of all things to be moist. Continuing his emphasis on the practical, Farrington speculated that Heraclitus chose fire as the first principle perhaps because it was the active agent which produced change in so many technical and natural processes. In marked contrast was Pythagorean society, in which contempt for manual labour kept pace with the growth of slavery and technical processes of production became more shameful, fit only for slaves. How fortunate and acceptable that the secret constitution of things was revealed not to those who manipulated nature but to the thinkers. This, Farrington speculated, marked the separation of philosophy from the techniques of production.[25]

The possible links between Presocratic cosmology and its social setting are not limited to Farrington's imagination. Jean Pierre Vernant set out to explain the change from arithmetical Babylonian astronomy to geometrical Greek cosmology by arguing from the general premise that social change preceded philosophical. Thus the rationalisation of science and cosmology followed the secularisation and rationalisation of political administration. He related this process to the reorganisation of social space within the city and the appearance of the open central public space, the agora, in Ionian and Greek cities. Consequently, he argued that cosmological space was reorganised when Anaximander placed the earth in the centre of the universe.[26] Such speculation seems to have quickly exhausted the few facts we think we know regarding the Presocratics.

While Farrington, the Marxist, argued that science and cosmology are derived from social structures and needs, others have searched for ways in which science and cosmology influenced society. The historian of science, Richard Olson, for instance, has sought out instances of the extension and application of scientific attitudes and modes of thought beyond the domain of natural phenomena to a wide range of cultural issues that involve human interactions and value structures. He concludes that the rise of Presocratic science and the intrusion of its 'attitudes and ideas into a collapsing intellectual structure accelerated the downfall of traditional beliefs, and was decisive in shaping and forming the religious, ideological, and moral traditions that replaced those grounded in Homer'.[27] The general thesis seems plausible enough, especially for modern times in which science and technology play an increasingly larger role in our lives, but a detailed and convincing articulation of the theme for the Presocratic period remains to be done.

Koestler's Sleepwalkers
Another innovative approach to Presocratic science flowed from the pen of the novelist Arthur Koestler into his book *The Sleepwalkers: A History of Man's Changing Vision of the Universe*. Koestler was interested in the psychological process of discovery and in the process that initially blinds a person towards truth which, once perceived, is regarded as heartbreakingly obvious.[28] He examined the unconscious biases and philosophical and political prejudices of astronomers and scientists more than a decade before the physicist and historian of science Gerald Holton coined the term *themata* to describe the underlying beliefs, values, and world views that lie behind the quasi-aesthetic choices that scientists make and which guide their leaps across the chasm between experience and basic principle.[29] No branch of science, Koestler asserted, whether ancient or modern, could claim freedom from metaphysical bias of one kind or another. Although the progress of science generally was regarded as a clean, rational advance along a straight, ascending line, in fact it zigged and zagged, so Koestler argued, nearly two decades before Thomas Kuhn questioned the common notion of scientific progress.[30] He saw that the history of cosmic theories, in particular, was a history of collective obsessions and controlled schizophrenias, more of a sleepwalker's performance than of an electronic brain.

Then, in Kepler's unfolding story, came Pythagoras, whose influence on the ideas of the human race was all-encompassing, uniting religion and science, mathematics and music. He took the first steps toward the mathematisation of human experience, the beginning of science. His emphasis was on form, proportion, and pattern, on the relation rather than on the relata. The Pythagorean dream that musical harmony governed the motion of the stars, though, more a dream dreamt through a mystic's ear than a working hypothesis, more a poetic conceit than a scientific concept, retained its mysterious impact, reverberating through the centuries and calling forth responses from the depth of the unconscious mind. In the sixteenth century Kepler, enamoured of the Pythagorean dream, used its foundation of fantasy to build, by equally unsound reasoning, the solid edifice of modern astronomy.

Scientific Highlights
Koestler attracted few followers, and interest in Presocratic cosmology moved back to the ideas themselves and to the argument that they represented increasing rationality on the road to modern science. Marshall Clagett, one of the first professional historians of science emerging from university studies in the United States after World War II, insisted in his *Greek Science in Antiquity* that the tone of much of Presocratic philosophy was rational, critical, often secular, and non-mythological. He argued that the critical spirit that emerged from this period was of great significance for the subsequent growth of science, especially the emergence of a theoretical and abstract science, in which sets of empirical rules were replaced by more generalised ones. Clagett did admit that the schemes of Thales and his successors originated in analogies and patently insufficient observational data. The Pythagoreans, on the other hand, used mathematics to deepen the ties between their theoretical explanation of nature on the one hand and their experience of nature on the other.[31] Clagett showed enthusiasm, sympathy, and understanding, and his book has yet to be displaced. It has, though, been characterised as the last of the old-style general handbooks, concentrating on science separately from its philosophical background, as a history of scientific highlights, rather than an attempt to understand both ancient science and the society that produced it.[32]

Source books constitute another category of scholarly text, and for Presocratic science there is *A Source Book in Greek Science*. The editors, Morris Cohen and I. E. Drabkin, a philosopher and a classicist, realised that it was an error to study the past exclusively from the point of view of current conceptions, judging ancient science according to modern criteria, but they also were concerned to discriminate between genuine science and folklore.[33] The resulting book has been criticised for looking at ancient science through modern quantitative spectacles, concentrating on only the highest and most successful examples, which happen to be mathematics, astronomy, and mathematical geography. The editors included only what they regarded as scientific material and omitted any reference to philosophical speculation. In any future source book more attention should be given to the intellectual background and how the ancients organised and systematised their own thinking about nature. The case that this should be so has already been convincingly made.[34]

Cohen and Drabkin's failure to consider the broader context of the development of quantifiable science is hardly new, and is characteristic of many previous histories. Thomas Kuhn, a critic of such work, described its methods. He wrote that it sought 'to clarify and deepen an understanding of contemporary scientific methods or concepts by displaying their evolution. Committed to such goals, the historian characteristically chose a single established science or branch of science - one whose status as sound knowledge could scarcely be doubted - and described when, where, and how the elements that in his day constituted its subject matter and presumptive method had come into being. Observations, laws, or theories which contemporary science had set aside as error or irrelevancy were seldom considered unless they pointed a methodological moral or explained a prolonged period of apparent sterility'. In Kuhn's words the scientist-historian viewed 'the development of science as a quasi-mechanical march of the intellect, the successive surrender of nature's secrets to sound methods skilfully deployed'. Only gradually have historians of science come 'to see their subject matter as something different from a chronology of accumulating positive achievement in a technical specialty defined by hindsight'.[35]

Precursor to Modern Science
One of the most positive appraisals of Milesian philosophy vis-à-vis modern science was written in the 1950s by S.Sambursky, a physicist. In *The Physical World of the Greeks* he purported to find striking similarities between patterns of thought in ancient Greek and modern science, and he presented 'noteworthy examples of the scientific approach that in the sixth century opened up a new era in the history of systematic thought...the teaching of the Milesian philosophers, which is remarkable for its rationalism'.[36] Sambursky was less enthusiastic about the Pythagoreans, though he did concede that their application of mathematics to basic physical phenomena conformed with correct modern method. Despite the similarity of formal scientific approach, however, Sambursky claimed an essential difference: the Pythagoreans extrapolated humanity into the cosmos, while modern science attempts to project mathematical and physical laws into man.[37]

Sambursky came to his study with a background in science and emphasised the emergence of modern science. Jacob Bronowski, on the other hand, came as a mathematician and, not surprisingly, focused his attention and his film 'The Music of the Spheres' (in *The Ascent of Man* series) on Pythagoras's search for a basic relation between mathematics and phenomena of nature. In his view progress in astronomy and physics followed from their amenability to mathematical treatment, and the laws of nature have been made of numbers since Pythagoras said number was the language of nature.[38]

Even more positive regarding the Pythagorean contribution to human advance were Olaf Pedersen, a historian of science, and his co-author Mogens Pihl, a physicist. In *Early Physics and Astronomy: A Historical Introduction,* they wrote that 'there is a mathematical structure behind the visible universe; the description of nature must therefore be expressed in terms of mathematics. From now on [after Pythagoras], this connection between physics and mathematics takes a progressively stronger hold upon the minds of natural philosophers, and must be thought of as the most important contribution to the advancement of science made by the Pythagoreans. It retained its fascination, and its inspiration to scientists persisted even after the specific Pythagorean doctrines had been abandoned as naïve or as obscure manifestations of an arbitrary number mysticism'.[39] This book was intended as an introductory textbook, and certain choices had to be made. Still, as one reviewer asked: 'Is it best to present the many physical concepts in a manner that is readily intelligible to the modern reader even if it means mathematising what was often rendered verbally and wrenching out of context ideas that may have been submerged in philosophical, metaphysical, and even theological discussion? And is it justifiable to concentrate on those aspects of ancient and medieval physical thought that adumbrated or heralded ideas and concepts that would prove significant in the scientific revolution, while ignoring by far the largest portion of early physical thought which might strike the modern reader as crude and irrelevant?'[40]

Rational Debate
Enthusiasm was expressed for the Milesians by the classicist G.E.R.Lloyd, though not because of any purported similarities with modern science. In *Early Greek Science: Thales to Aristotle,* Lloyd finds two major achievements of Milesian philosophy: the rejection of supernatural explanations of natural phenomena and the institution of the practice of rational criticism and debate. He saw that dogmatic though Presocratic philosophers were in presenting their answers, still they tackled the same problems, investigated the same natural phenomena, and were aware of the need to examine and assess their opponents' theories. Lloyd attributes this practice of debate to political conditions in Greece

and an extension of the customs of political debate to scientific inquiry.[41] 'But', he wrote, 'while philosophy and science did not involve a different mentality or a new logic, they may be represented as originating from the exceptional exposure, criticism and rejection of deep-seated beliefs...So far as an additional distinctively Greek factor is concerned, our most promising clue (to put it no more strongly) lies in the development of a particular social and political situation in ancient Greece, especially the experience of radical debate and confrontation in small-scale, face-to-face societies...(and) those who deployed evidence and argument were at an advantage...'[42]

Jonathan Barnes, too, in *The Presocratic Philosophers* emphasises the role of open debate in the development of cosmology and considers that 'What is significant is not that theology yielded to science or gods to natural forces, but rather that unargued fables were replaced by argued theory, that dogma gave way to reason...Few Presocratic opinions are true; fewer still are well grounded. For all that, they are, in a mild but significant sense, rational: they are characteristically supported by argument, buttressed by reasons, established upon evidence'.[43] Farrington earlier had argued from a different position in *Science and Politics in the Ancient World*. Far from political debate fostering science, he argued that scientific activity had declined in the ancient world when the struggle between science and obscurantism ultimately became a political struggle. He thought that scientific schools did save the Greeks from hierarchic petrification, but only temporarily. He drew attention to the threat Ionian philosophy posed to the institution of the state cult, and the Ionian philosopher Anaxagoras' (c.500 - c.428 BC) expulsion from Athens after his new theory of universal order posed a threat to the popular belief that celestial phenomena were controlled by the gods.[44] Actually, Anaxagoras was indicted both for impiety and for corresponding with agents of Persia, whose subject he formerly had been.[45] Some historians choose to believe that the jury which judged him responded at least in part to the avowed charges of impiety, while other scholars elect 'to emphasise immediate political reasons for the persecution and downplay the claimed science-impiety association as an incidental rationale, unimportant in itself'.[46]

Ludwig Edelstein, a classicist, medical historian, and philosopher, was not convinced that Farrington had sufficient evidence to uphold his argument on the interaction between science and politics. 'In every respect, then,' he wrote, 'Farrington's explanation of the development of ancient science seems to be untenable. His books have done much to arouse interest in the subject. The thesis which they advocate is vitiated however by what, in my opinion, is the basic error in many of the recent evaluations of ancient science, namely, the misapplication of historical analogies. Conditions in antiquity are seen in the light of subsequent events. The conflict between science and religion, which characterised later ages, is injected into the ancient world. Progress and decay of Greek and Roman science are judged by the standards of modern science.'[47] Whatever the merits of this criticism, which Farrington had invited upon himself by citing examples from modern times to support his contention that interaction between science and politics does take place, Farrington's interpretations of ancient science have failed to attract a significant following.

Many historians are open to the charge of overemphasising in the past problems of the present. Indeed, as Richard Olson has written, there exists 'near paranoia about the whiggishness of the history of science as a discipline. We seem to agree on almost nothing but the need to avoid imposing inappropriate modern categories upon historical activities, and the need to otherwise avoid reading the present into the past. Thus, we are almost apologetic about speaking of Greek science at all...'[48]

The Pythagoreans
Lloyd, while relatively enthusiastic about the Milesians, is only lukewarm when it comes to the Pythagoreans. The two philosophies were distinguished by their religious beliefs and cosmological theories. Granted, the Pythagoreans were the first to give knowledge of nature a quantitative, mathematical foundation, and hence could be considered scientific. Yet they held not only that phenomena are expressible in numbers, but also that things are made of numbers, this defying most modern conceptions of science. Furthermore, Lloyd concludes, 'many of the resemblances that the Pythagoreans claimed to find between things and numbers were quite fantastic and arbitrary'.[49] The Pythagoreans attracted few followers in ancient times, and Lloyd, who writes only about ancient science, rightly accords them scant attention. Those who look ahead to the Renaissance, and particularly to Kepler, though they need not attribute to the original Pythagoreans all the importance later followers achieved, cannot ignore the Pythagorean emphasis on number. At least, most cannot. James Coleman, a scientist-historian, was so distressed, however, with incorrect opinions, that when he reached Kepler in his *Early Theories of the Universe*, he could not bring himself to mention Pythagoras by name. 'Kepler, too,' he wrote, 'was a victim of

the fallacious reasoning of his predecessors, but even though Kepler was often forced to many years of fruitless labour because of convictions and philosophies about the universe which he inherited, he was quick to renounce not only the erroneous arguments of predecessors but his own follies as well when this path was indicated. The clearing away of the debris enabled Kepler, with his prodigious persistence, finally to be led to the first correct description of the seemingly haphazard motions of the planets.'[50]

This perception of Kepler, fighting free of evil Pythagorean influence rather than beneficently guided by it, along with the author's focus on correctness, enabled him to see that the Pythagoreans' main contribution lay not in using mathematics to increase ties between theoretical explanation of nature on the one hand, and experience of nature on the other, as have most writers on Presocratic science, but instead in their discovery that the earth is round. Also, Coleman repeatedly found it necessary to remind his readers that the Milesians' ideas were not correct: 'That Thales was incorrect' he wrote 'is obvious in the light of the relatively vast knowledge of today...Its importance lay not in the model itself, which today is known to be incorrect, but in the fact that Anaximander was the first person to reduce the workings of the universe to a mechanical system...The "model" itself was incorrect in the light of today's knowledge, but before the facts could be established a long chain of progressively correct interpretation of astronomical discoveries had to be established.'[51]

In contrast to those scientist-historians who are perhaps more enthusiastic than erudite, Thomas L.Heath is a respected scholar whose pioneering work on Greek science remains a valuable source. In terms of emphasis and interpretation, however, his major book on early Greek astronomy falls among the older histories of science since castigated by Kuhn as chronologies of accumulating positive achievement seldom considering observations, laws, and theories which contemporary science has set aside as erroneous or irrelevant.[52] Heath set himself the stated task of 'tracing every step in the progress toward the true Copernican theory' and showing 'that Aristarchus [not Heraclides of Pontus, as Giovanni Schiaparelli had asserted] was the real originator of the Copernican hypothesis'.[53] He looked primarily at those discoveries and observations validated as scientific by modern standards: 'Thales' claim to a place in the history of scientific astronomy depends almost entirely on one achievement attributed to him, that of predicting an eclipse of the sun,'[54] while 'Anaximander boldly maintained that the earth is in the centre of the universe...'[55] The first sentence of his chapter on Anaximenes began: 'For Anaximenes of Miletus...the earth is still flat...',[56] while he described Anaxagoras as 'A great man of science (who) enriched astronomy by one epoch-making discovery. This was nothing less than the discovery of the fact that the moon does not shine by its own light but receives its light from the sun. As a result, he was able to give (though not without an admixture of error) the true explanation of eclipses.'[57] Pythagoras is credited with his eponymous theorem, with inventing the science of acoustics, his discovery regarding musical tones, and a spherical earth, but there is not even a hint that he had anything to do with some mystical philosophy regarding a relationship between mathematics and phenomena of nature. The remarkable development by later Pythagoreans, in Heath's opinion, was their abandonment of the geocentric hypothesis.[58]

Thales' Eclipse Prediction
Thales' purported eclipse prediction marks for many scientist-historians the beginning of Western astronomical science. Retrospective astronomical calculations showing a total solar eclipse on 28 May 584 BC in Northern Turkey, help confirm Herodotus' report that 'In the sixth year of the war, which they [the Medes and the Lydians] had carried on with equal fortunes, an engagement took place in which it turned out that when the battle was in progress the day suddenly became night. This alteration of the day Thales the Milesian foretold to the Ionians, setting as its limit this year in which the change actually occurred'.[59] Presumably the warring parties either took the eclipse of the sun as a sign to cease fighting, or they were eager for any reason to cease and found the eclipse a convenient excuse. Most historical discussion has centred not on the credibility of the tradition itself, but on what methods Thales could have used to predict the solar eclipse. Willy Hartner has argued that Thales could have predicted an eclipse before the end of 583 BC from a study of the periodic recurrence of solar eclipses, and then taken credit for a different eclipse occurring slightly earlier.[60]

Thales' prediction of the solar eclipse of 584 BC may, however, be more myth than historic truth and as Alden Mosshammer has pointed out, 'As modern research in the history of ancient science and mathematics has advanced, confidence in Thales' ability to predict a solar eclipse has receded'.[61] Weighing in with the most caustic damnation of the credulity of his naive colleagues is Dicks, who found their conclusions 'totally at variance with the available evidence...of Thales' alleged prediction of a solar eclipse. In a desperate attempt to vindicate the historicity of this prediction, [the scholar] spins a web of inferential reasoning, based on wholly improbable suppositions...presupposing not only accurate observations, but also the concept of the ecliptic...the assumption that such comparatively advanced astronomical knowledge was possible in the sixth century BC is ludicrous; as we have seen, all the indications are...that such a stage was not reached until at least 150 years later'.[62]

Fresh Views
Providing a welcome contrast to the older-type histories is Stephen Toulmin and June Goodfield's *The Fabric of the Heavens*. He was trained as a physicist but later became a historian of science, and she was trained by him as a historian of science. In a series of books on the development of scientific thought, they set out 'to illustrate and document the manner in which our chief scientific ideas have been formed'.[63] This beginning could all too easily have led to yet another chronology of accumulating positive achievement, but they realised that 'to understand fully the scientific traditions which we have inherited, it is not enough to discover what our predecessors believed and leave it at that: we must try to see the world through their untutored eyes, recognize the problems which faced them, and so find out for ourselves why it was that their ideas were so different from our own...Different situations gave rise in earlier times to different practical demands; different practical demands posed different intellectual problems; and the solution of these problems called for systems of ideas which in some respects are not even comparable with our own'.[64] In other words the Presocratics did have some ideas which are now judged correct, but they did not elaborate, test them, or prove them. The union of theory and practice characteristic of modern science came later. Presocratic science was purely an intellectual enterprise undertaken with no technological end in view. For wild generalisation or unsound theorising or incautious analogy there was no potential penalty to pay in bridges collapsed or lives lost, and hence also no shackles on originality and imagination.[65] On the Pythagoreans, Toulmin and Goodfield argued that 'the most grandiose ambition they conceived was to explain all the properties of nature in arithmetical terms alone', and their 'belief that the distances of the planets from the centre of their orbits fit a simple "harmonious" mathematical law was the life-long conviction of Kepler, two thousand years later, and inspired the whole course of his astronomical researches'.[66]

The intellectual nature of Presocratic science and the separation of theory from practice are also themes in a joint appraisal of the

Pythagoreans by Bernard R.Goldstein and Alan C.Bowen, a historian of science and a classicist. They wrote that 'The Pythagoreans regarded the explanation of the heavenly motions in terms of these ratios as knowledge of the speeds, risings, and settings of the celestial bodies; and Plato called it astronomy. But, though such speculation did relate celestial movement and number, it would be wrong to see in this any attempt at precise measurement of what is observed. The explanandum in these theories is not so much a physical phenomenon as the ethical and aesthetic order it supposedly exhibits.'[67]

A possible explanation of the Presocratic attitude toward theories, especially the apparent lack of interest in testing them, focuses on their emphasis on problems of cosmogony (how the world came into being) rather than of cosmology (the current structure and future evolution of the world), and the consequent direction of their scientific efforts to the past than to the present and the future. 'As might have been expected in an age whose central problem was cosmogony, i.e., a set of unobservable and unrepeatable phenomena, and which, moreover, lacked all magnifying devices, the need for increased factual knowledge and for testing assumptions by experience was hardly felt. The facts to be explained were supposed to be matters of common knowledge, and any endeavour to account for them was essentially like the effort to solve a riddle. A scientific hypothesis was a (more or less fortunate) guess and the only criterion of its validity was its intrinsic plausibility.'[68] Lloyd, too, makes the point that much of the Presocratics' speculative effort was concentrated on astronomy, and though there might be attempts to verify theories with future observations, astronomy, strictly speaking, is not an experimental science, as it was impossible to vary or govern conditions of the objects under observation; direct experimentation was therefore impossible.[69]

Popper's Philosophy

The Presocratics may not have tested their theories, but did they discuss them? The matter of a tradition of critical discussion has been raised in a philosophical context by the philosopher of science Karl Popper. He asked wherein does the much discussed 'rationality' of the Presocratics lie? Not in any empiricism, because the Presocratics were critical and speculative rather than empirical. Yet when Popper wrote this, in the late 1950s, both traditional empiricist epistemology and traditional historiography of science were still, according to him, deeply influenced by the Baconian myth that science starts from observation and then slowly and cautiously proceeds to theories.[70] Science, according to this myth, began only when the speculative method was replaced by the observational method, when deduction was replaced by induction. For Popper, however, observations and experiments do not lead to an expansion of conjectural or hypothetical knowledge. Instead, observations and experiments play only the role of critical arguments, and their significance lies entirely in how they may be used to criticise theories.[71] From this theory of knowledge, it was but a short step for Popper to identify the modern rationalist tradition with the ancient Greek tradition of critical discussion. He identified the element of rationality in the Presocratics' thought in their attempt to know the world as the critical self-examination of their theories. Knowledge, Popper argued, proceeds by way of conjecture and refutation, and Presocratic philosophy developed through the clash of ideas in a critical debate.[72]

While pursuing his discussion of scientific methodology Popper somewhat inadvertently criticised Kirk's interpretation of Heraclitus. Kirk felt compelled to reply, upholding his interpretation of Heraclitus and also attempting to chip away at Popper's view of science.[73] Lloyd dismisses much of the squabble between Popper and Kirk as more of a difference between academic specialities than a disagreement over content. He also shows persuasively that broadening the question of scientific methodology to other fields of early Greek science produces different answers and rightly considers it a minor scandal that the debate initiated by Popper fizzled out so quickly.[74]

Following Popper, Lloyd further emphasised debate among the Presocratics in an article a few years later. 'Greek cosmology is nothing if not dialectical. And this is not an accidental or contingent feature of Greek cosmology, but of the essence of the Greek contribution.'[75] Greek cosmologists were in competition with each other for the best explanation, for the most adequate theory, and had 'an awareness of the need to examine and assess theories in the light of the grounds adduced for them...The history of early Greek cosmology is one of argument and counter-argument with a paucity of references to empirical data, and those mostly familiar ones'.[76]

Conclusion
Popper's philosophical emphasis has not won over classicists. Indeed, much of the disagreement over the nature of Presocratic cosmology can be understood in terms of the interests of different academic specialities and different assumptions about the nature of science. As Holton has

observed: 'The search for answers in the history of science is itself imbued with themata...we must be prepared for the criticisms of those who are afflicted, not with our themata, but with their antithemata.'[77] And from Lloyd: 'whether or not historians make explicit their views on the philosophy of science, the history they write will inevitably incorporate judgements, on the nature of science itself, on what demarcates it from other inquiries, on scientific methodology.'[78]

With a limited amount of raw material, each new thesis quickly exhausts inherent possibilities. Interpretations of Pythagorean and Milesian cosmology and culture have little chance of becoming paradigms for the practice of what we might term 'normal' history, in analogy to Thomas Kuhn's normal science, which finds practitioners in agreement upon certain basic problems and techniques and industriously expanding and elaborating an initial idea. Here, Lloyd's work may turn out to be a happy exception to the general absence of sustainable intellectual themes in studies of early Greek cosmology. Also, Lloyd avoids the extremes of both scientist-historian and classicist; he offers historical perspective tempered by critical sense, and shows the forest as well as the trees. It is no easier to imagine means for testing speculations about Presocratic cosmology and culture than it was for the Presocratics to test their own speculations. Caught up in an intellectual speculative fever, we must be cautious lest we become so entranced that we lose our footing.

References

1. Victor E. Thoren, review of D. R. Dicks, *Early Greek Astronomy to Aristotle,* (Ithaca, New York: Cornell University Press, 1970), in Isis, 61 (1970), 541-2.
2. William D. Stahlman, review of G. S. Kirk, ed., *Heraclitus, the Cosmic Fragments* (Cambridge: Cambridge University Press, 1954), in Isis, 45 (1954), 308-9. For disagreement among the trees, see Gregory Vlastos, 'On Heraclitus', *American Journal of Philology*, 76 (1955), 337-68; reprinted in David J. Furley and R. E. Allen, eds., Studies in Presocratic Philosophy. Vol. 1, *The Beginnings of Philosophy* (London: Routledge & Kegan Paul, 1970), pp. 413-29.
3. A very brief characterisation of the two philosophies has the Milesians driven by intellectual curiosity and dissatisfaction with the old mythological models to create a systematic natural explanation for physical and celestial phenomena, but their theories tended to be untestable and dogmatic, and if evidence clashed with dogma they preferred the dogma. The Pythagoreans were more concerned with metaphysical explanations and driven more by religious imperatives than scientific ones. The Pythagoreans were characterized by Aristotle: 'the Pythagoreans, as they are called, devoted themselves to mathematics; they were the first to advance this study, and having been brought up in it they thought its principles were the principles of all things...things seemed in their whole

nature to be modelled after numbers, and numbers seemed to be the first things in the whole of nature, they supposed the elements of numbers to be the elements of all things, and the whole heaven to be a musical scale and a number.' (*Metaphysics* I 5, 985b23-986a3) And Aristotle on the Milesians: 'Most of the first philosophers thought that principles in the form of matter were the only principles of all things: for the original source of all existing things, that from which a thing first comes-into-being and into which it is finally destroyed, the substance persisting but changing in its qualities, this they declare is the element and first principle of existing things, and for this reason they consider that there is no absolute coming-to-be or passing away, on the ground that such a nature is always preserved...for there must be some natural substance, either one or more than one, from which the other things come-into-being, while it is preserved. Over the number, however, and the form of this kind of principle they do not all agree; but Thales, the founder of this type of philosophy, says that it is water (and therefore declared that the earth is on water), perhaps taking this supposition from seeing the nurture of all things to be moist, and the warm itself coming-to-be from this and living by this (that from which they come-to-be being the principle of all things) - taking the supposition both from this and from the seeds of all things having a moist nature, water being the natural principle of moist things'(*Metaphysics* I 1, 983b6-27). An important book on Pythagoreanism is Walter Burkert's *Lore and Science in Ancient Pythagoreanism* (Harvard UP), 1972), which has a long section on astronomy. Also of note are the long chapter on astronomy in C.A.Huffman, *Philolaus of Croton* (Cambridge UP, 1993) and Maria Papathanassiou, 'The Influence of Pythagorean Philosophy on the Development of Mathematical Astronomy', in K.I.Boudouris (ed.), *Pythagorean Philosophy* (Athens, 1992). A good overall study is D.J.Furley, *The Greek Cosmologists*, Cambridge UP, 1987.

4. M. L. West, 'Alcman and Pythagoras', *Classical Quarterly*, 61 (new series 17), (1967), 1-15.

5. G. S. Kirk, J. E. Raven, and M. Schofield, *The Presocratic Philosophers: A Critical History with a Selection of Texts*, 2nd. ed. (Cambridge: Cambridge University Press, 1983). Astronomical and cosmological material from this volume has been abstracted in Norriss S. Hetherington, *Ancient Astronomy and Civilization* (Tucson, Arizona: Pachart, 1987); see also 'Early Greek Cosmology', in Hetherington, ed., *Encyclopedia of Cosmology: Historical, Philosophical, and Scientific Foundations of Modern Cosmology* (New York: Garland, 1993), pp. 183-8, and 'The Presocratics', in Hetherington, ed., *Cosmology: Historical, Literary, Philosophical, Religious, and Scientific Perspectives* (New York: Garland, 1993), pp. 53-66. For entry into the voluminous literature on the Presocratics, see the introduction to bibliographic tools in the editor's supplement (pp. xvii-xxvii) and selective bibliographies to 1974 (pp. 527-542) and from 1973 to 1993 (pp. xxix-xlvii) in Alexander P. D. Mourelatos, *The Pre-Socratics: A Collection of Critical Essays*, revised ed. (Princeton, New Jersey: Princeton University Press, 1993), and Luis E. Navia, *The Presocratic Philosophers: An Annotated Bibliography* (New York: Garland, 1993). The major bibliography is L. Paquet, M. Roussel, and Y. Lafrance, *Les Présocratiques: Bibliographie analytique* (1879-1980), 2 vols. (Montreal: Bellarmin, 1988-89). Also see R.D.McKirahan, *Philosophy Before Socrates*, Hackett, 1994.

6. Kirk, *Heraclitus* (ref. 2), p. 30. See especially H. F. Cherniss, *Aristotle's Criticism of Presocratic Philosophy* (Baltimore: Johns Hopkins University Press, 1935); summarized in 'The Characteristics and Effects of Presocratic Philosophy', *Journal of the History of Ideas*, 12 (1951), 319-45; reprinted in Furley and Allen, *Studies in Presocratic Philosophy* (ref. 2), pp. 1-28. Cherniss' thesis is criticized not as incorrect, but perhaps as going rather too far, in W. K. C. Guthrie, 'Aristotle as a Historian: Some Preliminaries', *Journal of*

Hellenic Studies, 77 (1957), 35-41; reprinted as 'Aristotle as a Historian', in Furley and Allen, *ibid.*, pp. 239-54. For a criticism, in turn, of Guthrie's article, see J. G. Stevenson, 'Aristotle as a Historian of Philosophy', *Journal of Hellenic Studies*, 94, (1974), 138-43. See also J. B. McDiarmid, 'Theophrastus on the Presocratic Causes', *Harvard Studies in Classical Philology*, 61 (1953), 85-156; reprinted, with abridgments, in Furley and Allen, ibid., pp. 178-238. An important new work, taking a fresh look at the Aristotelian view of the Presocratics, is Peter Kingsley, *Ancient Philosophy, Mystery and Magic*, Oxford UP, 1995.
7. Thoren, review of Dicks, *Early Greek Astronomy to Aristotle* (ref. 1).
8. D. R. Dicks, 'Solstices, Equinoxes, & the Presocratics', *Journal of Hellenic Studies*, 86 (1966), 26-40.
9. D. R. Dicks, 'Thales', *Classical Quarterly*, 53 (new series 9), (1959), 294-309.
10. Dicks, 'Solstices, Equinoxes, & the Presocratics', (ref 8).
11. Charles H. Kahn, 'On Early Greek Astronomy', *Journal of Hellenic Studies*, 90 (1970), 99-116. See also 'Some Remarks on the Origins of Greek Science and Philosophy', in Alan C. Bowen, ed., *Science and Philosophy in Classical Greece* (New York: Garland, 1991), pp. 1-10.
12. Jonathan Barnes, *The Presocratic Philosophers*, rev. ed. (London: Routledge & Kegan Paul, 1981), pp. 47-8; Heinrich Gomperz, 'Problems and Methods in Early Greek Science', *Journal of the History of Ideas*, 4 (1943), 61-76; reprinted in Daniel S. Robinson, ed., *Philosophical Studies by Heinrich Gomperz* (Boston: Christopher, 1953), pp. 72-87, and in Philip P. Wiener and Aaron Noland, eds, *Roots of Scientific Thought: A Cultural Perspective* (New York: Basic Books, 1957), pp. 23-38.
13. Barnes, ibid., p. 48.
14. G. E. R. Lloyd, 'Greek Cosmologies', in Carmen Blacker and Michael Loewe, *Ancient Cosmologies* (London: George Allen & Unwin, 1975), pp. 198-224. Reprinted, with an introduction assessing scholarly debate on the topic and Lloyd's modifications and developments in his own position since the original publication of the article, in Lloyd, *Methods and Problems in Greek Science* (Cambridge: Cambridge University Press, 1991), pp. 141-63.
15. Francis Macdonald Cornford, *From Religion to Philosophy: A Study in the Origins of Western Speculation*, 2nd ed. (Sussex: Harvester Press, 1983), pp. v-vii. Original edition, London: Edward Arnold, 1912.
16. Ibid., p. 143.
17. Ibid., p. 144.
18. Giorgio de Santillana, *The Origins of Scientific Thought: from Anaximander to Proclus 600 B.C.-500 A.D.* (Chicago: University of Chicago Press, 1961), p. 21.
19. Francis Macdonald Cornford, *Before and After Socrates* (Cambridge: Cambridge University Press, 1932), pp. 5, 7-8.
20. F. M. Cornford, 'Was the Ionian Philosophy Scientific?' Journal of Hellenic Studies, 62 (1942), 1-7; reprinted in Furley and Allen, *Studies in Presocratic Philosophy* (ref. 2), pp. 29-41. See also Cornford, *Principium Sapientiae: The Origins of Greek Philosophical Thought* (Cambridge: Cambridge University Press, 1952).
21. Gregory Vlastos, 'Cornford's Principium Sapientiae', *Gnomon*, 27 (1955), 65-76; reprinted in Furley and Allen, *Studies in Presocratic Philosophy* (ref. 2), pp. 42-55. Also see W.A.Heidel, *The Heroic Age of Science* (Baltimore, 1933) and Robin Waterfield, *Before Eureka, The Presocratics and their Science*, Bristol Press, 1989, Ch. 9
22. Benjamin Farrington, *Science in Antiquity*, 2nd ed. (Oxford: Oxford University Press, 1969), pp. 20-1, 27-8.

23. Stephen Toulmin and June Goodfield, *The Fabric of the Heavens* (London: Hutchinson, 1961), p. 64.
24. Bertrand Russell, *A History of Western Philosophy and Its Connections with Political and Social Circumstances from the Earliest Times to the Present Day* (New York: Simon and Schuster, 1945), p. 31.
25. Benjamin Farrington, *Greek Science* (Harmondsworth: Penguin Books, 1953), pp. 36-40, 48-9.
26. Jean Pierre Vernant, *Mythe et pensee chez les Grecs* (Paris: Libraire Francois Maspere, 1965); translated as *Myth and Thought among the Greeks* (London: Routledge & Kegan Paul, 1983), pp. 181-6, 190.
27. Richard Olson, *Science Deified & Science Defied: The Historical Significance of Science in Western Culture. vol 1. From the Bronze Age to the Beginnings of the Modern Era ca. 3500 B.C. to ca A.D. 1640* (Berkeley: University of California Press, 1982), pp. 62, 72.
28. Arthur Koestler, *The Sleepwalkers: A History of Man's Changing Vision of the Universe* (New York: Grosset & Dunlap, 1959), p. 14.
29. Gerald Holton, 'Themata in Scientific Thought', *in The Scientific Imagination: Case Studies* (Cambridge: Cambridge University Press, 1978), pp. 3-24. An earlier version of this essay, followed by commentary, appeared in Holton, 'Themata in Scientific Thought', *Science*, 188 (1975), 328-34, and Robert K. Merton, 'Thematic Analysis in Science: Notes on Holton's Concept', ibid., 335-8. On thematic analysis see also Holton, *Thematic Origins of Scientific Thought: Kepler to Einstein* (Cambridge, Massachusetts: Harvard University Press, 1973).
30. Thomas S. Kuhn, *The Structure of Scientific Revolutions*, 2nd ed., enlarged (Chicago: University of Chicago Press, 1970).
31. Marshall Clagett, *Greek Science in Antiquity* (London: Abelard-Schuman, 1955), pp. 34-35, 42, 43.
32. J. T. Vallance, 'Marshall Clagett's Greek Science in Antiquity: Thirty-five Years Later', *Isis*, 81 (1990), 713-21.
33. Morris R. Cohen and I. E. Drabkin, eds., *A Source Book in Greek Science*, 2nd ed. (Cambridge, Massachusetts: Harvard University Press, 1958), p. vii. See also Cohen, *A Dream's Journey: The Autobiography of Morris Raphael Cohen* (Boston: Beacon Press, 1949), p. 193.
34. Vallance, 'Marshall Clagett's Greek Science in Antiquity ', (ref. 32).
35. Thomas S. Kuhn, 'The History of Science', in *International Encyclopedia of the Social Sciences*, vol. 14 (New York: Crowell Collier and Macmillan, 1968), pp. 74-83; reprinted in Kuhn, *The Essential Tension: Selected Studies in Scientific Tradition and Change* (Chicago: University of Chicago Press, 1977), pp. 105-26.
36. S. Sambursky, *The Physical World of the Greeks* (London: Routledge and Kegan Paul, 1956), pp. 4-5. Translated by Merton Dagut from the Hebrew edition, *Kosmos shel ha-Yevanim* (Jerusalem: Bialik Institute, 1954).
37. Ibid., pp. 40, 42.
38. Jacob Bronowski, 'Music of the Spheres', 52-minute color film in *The Ascent of Man* series, no. 5 (BBC-TV and Time-Life Films, 1973, 16 mm. and videotape). See also Bronowski, *The Ascent of Man* (Boston: Little, Brown and Company, 1973), p. 187.
39. Olaf Pedersen and Mogens Pihl, *Early Physics and Astronomy: A Historical Introduction*, 2nd ed. (Cambridge: Cambridge University Press, 1994), pp. 17, 20.
40. Edward Grant, 'Physical Sciences before the Renaissance', *Journal for the History of Astronomy*, 7 (1976), 201-204. Grant's review was of the 1st edition of Pederson and Pihl.

41. G. E. R. Lloyd, *Early Greek Science: Thales to Aristotle* (London: Chatto & Windus, 1970), pp. 13-4.
42. G. E. R. Lloyd, *Magic, Reason, and Experience: Studies in the Origin and Development of Greek Science* (Cambridge: Cambridge University Press, 1979), pp. 265-6.
43. Barnes, *The Presocratic Philosophers* (ref. 12), p. 45.
44. Benjamin Farrington, *Science and Politics in the Ancient World* (London: George Allen & Unwin, 1939), pp, 70-1, 74-6.
45. T. W. Africa, *Science and the State in Greece and Rome* (New York: John Wiley & Sons, 1968), p. 39.
46. Richard Olson, 'Science, Scientism and Anti-Science in Hellenic Athens: A New Whig Interpretation', *History of Science*, 14 (1978), 179-99; Olson, *Science Deified & Science Defied* (ref. 27), pp. 79-82.
47. Ludwig Edelstein, 'Recent Trends in the Interpretation of Ancient Science', *Journal of the History of Ideas*, 1952, 13:573-604; reprinted in Wiener and Noland, eds., *Roots of Scientific Thought* (ref. 12), pp. 90-121, and in Owsei Temkin and C. Lilian Temkin, eds., *Ancient Medicine: Selected Papers of Ludwig Edelstein* (Baltimore: Johns Hopkins Press, 1967), pp. 401-39.
48. Richard Olson, 'Science, Scientism and Anti-Science', p 179-199. On whiggism, priggism, presentism, contextualism, and anti-antiwhiggism, see Stephen G. Brush, 'Scientists as Historians', *Osiris*, 10 (1995), 215-231.
49. Lloyd, *Early Greek Science* (ref. 41), pp. 24-6.
50. James A. Coleman, *Early Theories of the Universe* (New York: New American Library, 1967), p. 106.
51. Ibid., pp. 18, 21, 22.
52. Kuhn, 'The History of Science', (ref. 35).
53. Thomas Heath, *Aristarchus of Samos: The Ancient Copernicus. A History of Greek Astronomy to Aristarchus together with Aristarchus's Treatise on the Sizes and Distances of the Sun and Moon. A New Greek Text with Translation and Notes* (Oxford: Clarendon Press, 1913), p. iv. See also Heath, *Greek Astronomy* (London: J. M. Dent & Sons, 1932).
54. Heath, *Aristarchus of Samos*, p. 13.
55. Ibid., p. 24.
56. Ibid., p. 40.
57. Ibid., p. 78.
58. Ibid., pp. 46, 48, 94.
59. Herodotus, I, 74. See Kirk, Raven, and Schofield, *The Presocratic Philosophers* (ref. 5), pp. 81-2.
60. Willy Hartner, 'Eclipse Periods and Thales' Prediction of a Solar Eclipse - Historic Truth and Modern Myth', *Centaurus*, 14 (1969), 60-71.
61. Alden A. Mosshammer, 'Thales' Eclipse', *Transactions of the American Philological Association*, 111 (1981), 145-55. See also Otto Neugebauer, *Exact Sciences in Antiquity*, 2nd ed. (Providence, Rhode Island: Brown University Press, 1957), pp. 142-3.
62. Dicks, *Early Greek Astronomy* (ref. 1), p. 174.
63. Toulmin and Goodfield, *The Fabric of the Heavens* (ref. 23), p. 15.
64. Ibid., p. 16.
65. Ibid., pp. 68-9.
66. Ibid., pp. 79, 82.

67. Bernard R. Goldstein and Alan C. Bowen, 'A New View of Early Greek Astronomy', Isis, 74 (1983), 330-40; reprinted in Goldstein, *Theory and Observation in Ancient and Medieval Astronomy* (London: Variorum Reprints, 1985), pp. 1-11.
68. Gomperz, 'Problems and Methods in Early Greek Science', (ref. 12), pp. 31-2.
69. G. E. R. Lloyd, 'Experiment in Early Greek Philosophy and Medicine', *Proceedings of the Cambridge Philological Society*, 190 (new series 10), (1964), 50-72; reprinted, with an introduction assessing scholarly debate on the topic and Lloyd's modifications and developments in his own position since the original publication of the article, in Lloyd, *Methods and Problems in Greek Science* (ref. 44), pp. 70-99.
70. Karl R. Popper, 'Back to the Presocratics', Procedings of the Aristotelian Society, 59 (1958-1959), 1-24; reprinted, with additions, in Popper, *Conjectures and Refutations: The Growth of Scientific Knowledge* (London: Routledge and Kegan Paul, 1963), pp. 136-65, and in Furley and Allen, *Studies in Presocratic Philosophy* (ref. 2), pp. 130-153.
71. Ibid., on p. 151.
72. Ibid., on pp. 148-152.
73. G. S. Kirk, 'Popper on Science and the Presocratics', Mind, 69 (1960), 318-39; reprinted in Furley and Allen, *Studies in Presocratic Philosophy*, (ref. 2), pp. 154-77.
74. G. E. R. Lloyd, 'Popper versus Kirk: a Controversy in the Interpretation of Greek Science', *British Journal for the Philosophy of Science*, 18 (19670, 21-38; reprinted, with an introduction assessing scholarly debate on the topic and Lloyd's modifications and developments in his own position since the original publication of the article, in Lloyd, *Methods and Problems in Greek Science* (ref. 14), pp. 100-20, esp, p 105.
75. Lloyd, 'Greek Cosmologies', (ref. 14).
76. Ibid., pp. 209, 218-9.
77. Holton, 'Themata in Scientific Thought', (ref. 29).
78. Lloyd, 'Popper versus Kirk', (ref. 74), p. 100.

Changes in Celestial Journey Literature: 1400-1650

Alan S.Weber*

Introduction

This study investigates an important historical phase in the curiously hybrid genre of the celestial journey narrative which has produced not only important scientific texts, such as Macrobius's *Somnium Scipionis*, but also some of Western Europe's finest poems, including Dante's *Divine Comedy*. I would like to compare Christine de Pizan's *Chemin de Long Estude* of 1403, which describes the author's celestial journey through the heavenly spheres, to another milestone in celestial voyage literature, Francis Godwin's English work *The Man in the Moone* of 1638. These two literary and historical endpoints illustrate the changes in European technical astronomy which occurred between 1400 and 1650, and also reveal the shift which occurred in the very nature of the celestial voyage genre. I will also briefly review other closely related early modern celestial voyage narratives written by Johannes Kepler and Bishop John Wilkins.

The True History of Lucian of Samosata (born circa 117 C.E.), the satirical story of the author's trip to the moon, is often cited as a seminal text in the tradition of celestial journey literature because it contains one central element common to all narratives of this type: expansiveness. I mean here expansiveness in all its senses; Lucian undertakes the ultimate journey, past the pillars of Heracles, the limit of the known classical world, and this breaking of boundaries allows him to overstep other restrictions on etiquette, mores, and good taste. This expansiveness also allows him to satirize Homer, Socrates, Heracles, Aristophanes, and Dionysus with impunity.

Similarly, the breadth of the genre allowed an early medieval textbook writer, Macrobius (circa late fourth century C.E.) to use his celestial wanderings through the planetary spheres as the pretext for a journey into the secrets of Pythagorean number symbolism and the cosmology of Plato's *Timaeus*. One of the great poems of celestial journeying, Dante's *Divine Comedy*, closely resembles both Lucian and Macrobius in its use of the movement metaphor to explore, magnify, and expatiate upon unknown physical and epistemological territory. As he physically encompassed vast spaces, Dante also explored great ethical, spiritual, scientific, and theological truths, presenting his poem as an encyclopedia of both theology and cosmology. Dante was the direct inspiration for Christine de Pizan's celestial voyage in the *Chemin de Long Estude;* she certainly drew her encyclopedic purpose from him. Her poem therefore represents an excellent mirror of the cosmological thought of her day as she interpreted it within a tradition well known to her.

Christine de Pizan's Cosmology
The life and works of Christine de Pizan (1364-c.1430) have recently received renewed attention as evidenced by the numerous translations and critical studies of her works as well as the international conferences held in her honor during the last ten years. Christine was born in Venice in 1364, the daughter of Thomas of Pizan, who later became court astrologer to Charles V of France. Thomas of Pizan [or Bologna] taught astrology at Bologna from 1345 to 1356. He is well known for the magical charm he used to expel the invading English armies from French soil by burying wax images of the English commanders at various locations throughout the French kingdom.

Christine grew up in the learned atmosphere of the court of the French King Charles V. She began to write extensively in the 1390s after the death of her husband Etienne de Castel. Along with a biography of King Charles V, she produced a large corpus of short lyrics, longer narrative poems, and a series of didactic works on the position of women in 14th Century society. She became well known to her contemporaries through her debate with Jean de Montreuil, Gonthier Col, and Jean Gerson over the alleged obscenity and misogyny of Jean de Meun's popular poem *Le Roman de la Rose*.

The recent recovery of one of Christine de Pizan's neglected works, *Le Livre du Chemin de Long Estude* (1403), has involved the steady critical appreciation of a dream vision dismissed in the last century by Gaston Paris as a mediocre text, with some potential value, however, for the history of ideas.[1] The poem opens with Christine's ruminations on her recently deceased husband and the vicissitudes of *Fortuna*. She then ponders the sources of evil and change in the world, attributing mutation to the battle of the elements: 'fire and water hate each other/ And one desires to destroy the other'.[2] This is probably an example of the type of passage which Gaston Paris believed important for the history of ideas.
From her descriptions of the heavens in the opening sections of the poem, it is abundantly clear that Christine endorses the standard Aristotelian-

Ptolemaic model of the universe, so called because of its fusion of the fundamental principles of Aristotelian physics with Ptolemy's system of epicycles and geometrical determination of planetary orbits. Although the Aristotelian-Ptolemaic model remained the standard orthodoxy in western European universities during Christine's day, there were serious challenges to it throughout the Middle Ages, including such works as Bernardus de Silvestris's *Cosmographia*, which drew on Stoic, Hermetic, and Platonic thought.[3] It would be too simplistic, however, to divide medieval cosmology into Aristotelian and non-Aristotelian camps. Even astrology, which has often been portrayed in modern historiography as the antithesis of both Aristotelian cosmology and modern mathematical astronomy, was firmly rooted in the physical principles expounded in Aristotle's *De Caelo*, *De Generatione et Corruptione*, and the *Meteorologica*. As Richard Lemay observes, 'there can be little doubt that [Aristotle's works] supplied the scientific background of astrology during 2000 years after him'.[4] As we shall soon see, the European cosmology of Christine's age also eclectically adopted Stoic and Platonic ideas enriched by Arabic interpretations of Greek astronomy. Thus, European cosmology of the early modern period should not be characterized as slavish scholastic repetition of Aristotelian science, but by a richness of perspective, and Christine's interpretation of celestial science, which defines astronomy as a medium or bridge between divine and human knowledge, should not be lightly dismissed.

The greatest challenge to Aristotle's cosmological system ironically came from scholastic Christian theology itself, as evidenced by Bishop Etienne Tempier's condemnation in 1277 of 219 scholastic theses about the universe. Christine, however, does not mention any of the serious cosmological incompatibilities and tensions between Christianity and Aristotle, such as the question of the eternity of the world or the debate over the plurality of worlds.[5] Christine 'never ceases to consider Aristotle as the prince of philosophers' and liberally quotes extracts from Thomas Aquinas's commentary on Aristotle's *Metaphysics* in *Le Livre des Fais et Bonnes Meurs du Sage Roy Charles V*.[6] The Metaphysics also informs a large part of book two of *L'Avision Christine*, which may represent, as Glenda McLeod suggests, the first known vernacular commentary on Aristotle's *Metaphysics*.[7]

To return to Christine's narrative, she falls asleep during her reading and is visited by Sebille (Sybil) who announces that she will tell Christine the secrets of the universe because Christine is ready to conceive great knowledge ('apprestee a concevoir').[8] The double-entendre implicit in conceiving knowledge points to an important aspect of Christine's conception of science: she conceives of knowledge not as a body of facts brought from the outside world into the mind, but as an internal process of spiritualization, a birth of inward light.

Sebille next leads Christine to the fountain of Sapience, where all the great philosophers have drunk, including Socrates, the Cynics, Plato, Hermes, Seneca and her father Thomas of Pizan. From the fountain they proceed to the road of long study, a path well known to Christine. After whisking Christine around the various kingdoms of the world, Sebille leads her upwards to a high mountain, where she hears a clamour of Greek voices, no doubt the arguments of Stoics, Epicureans, Platonists, Skeptics, and Peripatetic philosophers all advancing their specific world views. Sebille next invites Christine to mount a subtle ladder extending from the heavens:

> Light it was and portable
> As if one could twist it around
> And carry it without labour
> Everywhere, if you wanted.
> You would never be hindered or bothered by it.
> It was not made of rope
> Nor any kind of cord or wood.
> I could not determine the material
> But it was long, strong, and light.[9]

Sebille calls the curious and resplendent device 'l'eschiele de Speculacion' - the ladder of speculation - a shining entranceway into the heavens ordained for those, who like Christine, love subtlety and learning. Now begins her ascent into the celestial spheres offering her the occasion to detail the structure of the heavens for the reader.

Christine and Sebille first pass through the first heaven ('premier ciel'), which is made out of air. From there, they pass into the second heaven, the 'ether', characterized by its shining brilliance and clarity. Instead of the common Aristotelian definition of the ether as the fifth element (*quinta essentia*) comprising the stars and existing only beyond the sphere of the moon, Christine accepts the Stoic conception of the ether as the purer part of the upper air. They next continue on into the third heaven or the sphere of fire, traditionally located under the sphere of the moon in Aristotelian cosmology. They continue on into the fourth heaven, called 'Olimpe', meant to recall Mount Olympus, the seat of the pagan gods and a general epithet in Old French for sky or heaven. Christine's ladder finally ends within the fifth heaven, 'le firmament', a place of pervasive and blinding light.

It is here, in the fifth heaven, that Christine receives a lesson in both astronomy, the science of the movements of the fixed and wandering stars (planets), and in astrology, the knowledge of the powers and influences exerted by those celestial bodies:[10]

> [Sebille] showed me everything, and told me the names and powers of the planets, and she made every effort to teach me the courses of the moving stars, both the fixed and the wandering. And she told me the properties, the effects, the contraries, the powers and the influences, and their various arrangements.[11]

After her description of the physical structure of the heavens, Christine curiously admits her inability to provide the reader with any further details, because she had not learned astrology at school.[12] Undoubtedly she knew of the technical treatises in the library of her astrologer father Thomas of Pizan, who must have possessed a collection of astrological texts as well as instruments, but how far she proceeded in her studies cannot be determined with any accuracy.

Both Edgar Laird and Charity Cannon Willard have demonstrated how astrology formed one of the central concerns of the learned court, which included Thomas, assembled by le Sage Roy Charles V.[13] The court of Charles V was very much a centre of scientific learning, especially of Aristotelianism. Charles had commissioned Nicole Oresme to translate the works of Aristotle into the vernacular. Oresme (c.1320 - 1382) was a Professor of Theology at the College of Navarre in Paris. He made great contributions to mathematics and physical science, and wrote several cosmological works on astronomy (*De l'Espere*), divination *(Livre de Divinacions)*, and astrology *(Tractatus Contra Judicarios Astronomos)*. The King was also a great patron of astrologers and founded the 'Collège de Maître Gervais' at the University of Paris for the study of astrology. Lynn Thorndike remarks about the court of Charles V: 'at this period wisdom and astrology were considered almost synonymous',[14] a viewpoint that constantly surfaces in Christine's works.

One source of Christine's astrological learning may have been one of the texts which went by the name of *On the Sphere*, such as Nicole Oresme's *De l'Espere* (late 14th Century).[15] Christine's description of the heavens echoes *De l'Espere* in several respects: both authors characterize the earth as a round ball as viewed from the moon. Both Christine and Oresme divide the heavens into five regions (in Oresme, three regions of the air, the sphere of fire, and the heavens). Oresme also mentions that Mount Olympus reaches into the upper air and Christine names one of her heavens 'Olimpe' in the *Chemin de Long Estude*.

Christine also makes reference in her poem to a wide variety of eclectic cosmological doctrines such as the planetary houses, the music of the spheres, and cosmic plenitude. She observes on her journey the houses of the planets ('les maisions que planetes ont'), and in which houses they are exalted ('quelles ont exaltacion').[16] She also describes the music of the spheres, the ordered movements of the heavens comprising 'l'armonie et belle chancon', a Platonic and Pythagorean idea widely disseminated in medieval literature despite Aristotle's skeptical rejection of the doctrine.[17] Her descriptions of the great number of heavenly bodies - 'la grant quantité pleniere/ Qui y est'[18] - reinforces the Aristotelian, Stoic, and later Christian idea of the plenitude of the cosmos. Scholastic Christian theology generally endorsed the idea of the Aristotelian and Stoic *plenum*, the absence of any empty space in the universe, in response to the *inane* or *kenon* (void space) of Epicurean physics which had denied divine providence and ordered causality. Void space implied a location where God was not, an affront to the creator's omnipresence. It is Christine's wonder at the plenitude of created nature which invites her to make her explorations through the celestial spheres.

Christine, however, finds that she cannot enter the Crystalline Heavens to see the nine orders of angels because of the present state of her corporeal body.[19] Sebille and Christine therefore descend to the sphere of the air, near the ethereal layer. There she meets the servants ('maigniee') of the 'intelligences haultaines,' the followers of the planets, sun, moon and other intelligences who are called Influences and Destinies. These Influences and Destinies are beings attached to every planet, intelligence, star, and heaven who serve them like household retainers.[20] Although she does not describe these beings at any length, they obviously act as mediums, messengers, and conduits between the divine powers of the heavenly bodies and material bodies on earth. In this section Christine seems to argue for complete astrological determinism, that our fates have been predestined by heavenly confluence. Yet she does remind her readers that God still rules the destinies from above:

> ...as soon as a man or a woman is born, however great, the destinies control their lives and assign them their proper end, good or evil,

according to the domain of the course of the planets at the hour when the infant is born. But nevertheless God, who has given them this power, reigns above and takes care of what pleases Him.[21]

Among the Destinies, Christine sees the rebellions, treasons, destroyed towns, and tempests of the evil fortunes about to be rained down on earth by the Influences. She even boasts of some prophetic knowledge: she says she now knows what the effects of the 1401 comet will be; but these effects, which she refuses to reveal to the reader, will not unfold for another 20 years.[22] The latter part of the *Chemin de Long Estude* consists of a court held by Queen Reason in which the four estates (Sagesse, Richesse, Chivalrie, and Noblesse) debate the proper virtues pertaining to the prince. Although to modern readers the second part of the poem may seem like a mirror for magistrates, or a conduct book such as Christine's *Treasure of the City of Ladies*, tacked on to an exposition of natural philosophy, I will argue later, after a description of Francis Godwin's work, for the essential unity of the poem. The *jugement* genre, highly developed by Machaut,[23] in which a central question is debated in a real or mock court by learned advocates, dovetails perfectly with Christine's encyclopedic introductory section of the *Chemin de Long Estude* since the *jugement* allows for the airing of diverse opinions, just as the tradition of commentary on Aristotle weighed and synthesized conflicting propositions on the nature of the cosmos. As Barbara K. Altmann aptly puts it: 'what more suitable forum for intricate argument than a court-room scene, where a minimum of plot could supply a certain amount of tension as to outcome and where characters of allegorical or human nature could quite justifiably defend conflicting opinions at length?'[24]

Godwin, Kepler and Wilkins
Separated in culture and time by over 200 years, Francis Godwin's *Man in the Moone*[25] (published 1638) nevertheless stands in direct line with Christine's *Chemin de Long Estude*; both works form part of the larger tradition of celestial journeys established by Lucian, Macrobius, and Dante.[26] Bishop Francis Godwin is best known for his *Catalogue of the Bishops of England since the first planting of Christian Religion in the Island, etc.*, for which he was awarded the bishopric of Llandaff by Queen Elizabeth.[27] He was one of a growing number of English clergymen, including Bishop John Wilkins, who took a serious interest in scientific matters and who attempted to reconcile astronomical advancements with divine writ. Godwin was writing after the detailed

observational work of Tycho Brahe, later published by Johannes Kepler, and the *De Revolutionibus* (1543) of Copernicus, which had proposed a heliocentric universe and rotating earth while retaining the fundamental Ptolemaic principles of circular motion and epicycles. After Tycho's parallax measurements of the 1577 comet had situated this extraordinary cosmic event well beyond the lunar sphere - in other words a decaying, changing object had appeared in Aristotle's alleged realm of perfection - Aristotelian cosmology, already damaged by Copernicanism, began seriously to unravel.[28] Godwin was also writing after the telescopic discoveries of Galileo reported in the *Siderius Nuncius* (1611).

In Godwin's work Domingo Gonsales, a Spanish Hidalgo, narrates his remarkable life story. After killing a man in a duel, Gonsales departs on a series of sea voyages which finally land him on the island of St. Helena. He discovers huge geese there - gansas - which he trains to carry heavy loads. He invents a flying machine by tying the gansas together, eventually flies off the island and is picked up by a Spanish ship. He convinces the captain to take the birds aboard, and soon after, the Spanish fleet is defeated by the English navy. Gonsales escapes with his gansas who inexplicably fly straight upwards into the atmosphere towards the moon. On his lunar voyage he encounters pleasing shapes floating in the air who bring him delicious foods. When he reaches the moon, on September 21, 1599, he finds that the food given to him by the aerial beings has turned to 'a mingle mangle of dry leaves, of *Goats haire, Sheepe*, or *Goats-dung, Mosse*, and such like trash'.[29] He has been deceived by wicked spirits.

As Gonsales arrives on the moon, we begin to see Godwin's somewhat transparent purpose in writing this narrative - he uses Gonsales to advance contemporary cosmological doctrines. In a similar fashion, Christine's journey had allowed her to provide an eyewitness account of the true order of the heavens. First, Gonsales refutes the existence of the

Gonsales en route to the Moon,
from Godwin's *The Man in the Moone* (1638)

Aristotelian sphere of fire, a ring of elemental fire attached to the upper air. Gonsales discovers that the upper air is of the same temperature as that below: 'Who is there that hath not hitherto believed the uppermost Region of the Ayre to be extreme hot, as being next forsooth unto the naturall place of the Element of Fire. O Vanities, fansies, Dreames!'[30]

We remember in Christine's poem how she fears the increasing heat as she mounts the ladder of speculation towards the Aristotelian sphere of fire.[31] She recalls the arrogance of Icarus who flew too high and presumptuously into the fiery heavens. Godwin, on the other hand, following common practice among professional astronomers of his age, has rejected the existence of the sphere of fire.[32] By the beginning of the seventeenth century, Stoic and Hermetic monism (universally operating physical law) began to supersede Platonic and Aristotelian dualism in physics (separate physical laws for heaven and earth), a development which challenged the doctrine of the elemental spheres.

Godwin also questions the Aristotelian idea of the natural place of the elements which made the centre of the earth the point towards which the heavier elements earth and water were attracted. In Godwin's cosmological system, magnetic force, an idea previously outlined in William Gilbert's widely read *De Magnete* (1600), replaces the natural place of the earth as a centre of attraction. This modification allows for the moon to possess a centre of attraction,[33] which Gonsales discovers as he walks on the moon. The lesser magnetic force of the moon allows him to bound high into the air.

Most importantly, Gonsales watches the earth turning on its axis from his position on the moon. He must therefore conclude along with the Copernicans that the earth spins on its axis every twenty four hours from west to east. We are very far away from Christine's fixed earth which she sees from above, sitting 'like a little, round ball'.[34] Gonsales now realizes the blindness of the philosophers who posit two contrary motions for the heavens and the planets:

> *Philosophers* and *Mathematicians* I would should now confesse the wilfulnesse of their own blindnesse. They have made the world believe hitherto, that the Earth hath no motion. And to make that good, they are fain to attribute unto all and every of the celestiall bodies two motions, quite contrary each to other; whereof one is from the *East* to the *West*, to be performed in 24 hours; (that they imagine to be forced, *per raptum primi Mobilis*) the other from the West to the East in severall proportions.[35]

Godwin and Gonsales, however, do not fully accept the Copernican hypothesis. Gonsales states:

> I will not go so far as *Copernicus*, that maketh the Sunne the Center of the Earth, and unmoveable, neither will I define any thing one way or other. Only this I say, allow the Earth his motion (which these eyes of mine can testifie to be his due) and these absurdities are quite taken away, every one having his single and proper Motion onely.[36]

Godwin's work has many affinities with Johannes Kepler's *Somnium*, a work on lunar astronomy published in 1634. With the aid of his mother, Kepler, who appears in the work as the character Duracotus, summons a demon who explains life on his home planet of Levania (the moon). Kepler uses this fiction, based on Lucian's *True History*, much in the same way as Godwin - to explicate lunar astronomy by shifting observational reference points from the earth to the moon. Thus, in his highly technical notes to the *Somnium*, Kepler explains eclipses, solstices, and lunar and terrestrial rotation as measured from the moon as a central reference point. Just as Godwin employed the fiction of the celestial journey to argue for terrestrial rotation, so Kepler states in his notes: 'here is the thesis of the whole *Dream*; that is, an argument in favour of the motion of the earth or rather a refutation of the argument, based on sense perception, against the motion of the earth'.[37]

John Wilkins' *The Discovery of A World in the Moone* was published in 1638, the same years as Godwin's *Man in the Moone* and four years after Kepler's *Somnium*. Wilkins was instrumental in founding the Royal Society and encouraged the study of astronomy at Oxford and London.[38] Unlike Godwin, Wilkins includes a full-blown assault on Aristotelianism. *The Discovery* clearly shows in what direction the genre of the celestial journey has travelled since the writing of Christine's *Chemin de Long Estude*. Wilkins altogether dispenses with the fictional narrative and presents a series of propositions about lunar and cosmic science. From the literary narrative revelatory of *scientia*, of which Christine de Pizan's work is a prime example, we have moved to the precursor of the scientific paper. As Marjorie Hope Nicolson points out: '[Wilkins's] *Discovery* is one of the first important books of modern "popular science", a work written by a man who knew the technicalities of science.'[39] Wilkins clearly sees himself working within the literary tradition of Lucian, Plutarch, Kepler, and Godwin and draws on these works, as well as

Galileo's *Siderius Nuncius*, for his lunar science. Yet Wilkins also concludes that a list of numbered propositions supported by evidence, in contrast to literary narrative, can best convey scientific knowledge to a literate audience.

Wilkins fully accepts the Copernican hypothesis (heliocentrism and terrestrial rotation) and also refutes some fundamental tenets of Aristotelian physics. First, Wilkins entertains the possibility of the plurality of worlds rejected by Aristotle, an idea according to Wilkins, that 'doth not contradict any principle of reason'.[40] Along with Godwin, he rejects the orb of elemental fire.[41] Proposition number three, which clearly demonstrates his break with Aristotelianism, boldly states: 'that the heavens doe not consist of any such pure matter which can priveledge them from the like change and corruption, as these inferiour bodies are liable unto'.[42] Wilkins has swept aside the barrier of elemental matter, which Aristotle had located at the sphere of the moon. The heavens for him can no longer consist of incorruptible, immaterial ether, but must be made of something more readily accessible to reason and *experimentum*.

Not only do we see a shift in specific astronomical doctrines in European cosmology from the time of Christine de Pizan to the work of Godwin and Wilkins, but also a profound difference in how the cosmos was perceived. Many of the technical changes in cosmology can be summed up simply as the growing rejection of Aristotelian physics. The book of *Genesis* and the Hexameral treatises (works combining theological and physical speculation on the first six days of creation) stood at the heart of early medieval cosmology before the Latin translations of Arabic and Greek astronomical texts entered the West. One book, the Bible, provided a coherent theology, physics, and cosmogony. By Wilkins's day, however, theology and biblical exegesis were becoming increasingly irrelevant to both astronomy and astrology, as practitioners of these sciences, especially in astronomy, were focussing their work more on practical computation, geometry, and mathematics than on origins, causes, and metaphysical powers of the Deity. As Wilkins points out regarding extrapolating the nature of the world from divine writ: 'such...absurdities [about nature] have followed, when men looke for the grounds of Philosophie in the words of Scripture'.[43]

Conclusion

For Christine de Pizan, astrology and astronomy were merely subsciences of theology - astrology represented that speculative ladder which mounts into the stars and reveals the high majesty. Theology in Christine's *Livre de la Mutacion de Fortune* is 'that supercelestial science which comprises all other knowledge'.⁴⁴ The study of Aristotelian metaphysics is likewise only a stepping stone - or ladder - to divine knowledge. We see similar statements regarding the proper place of astronomical and astrological learning within French culture in the work of Christine's contemporary Nicole Oresme. Oresme writes:

> ...astrology has three very noble ends. The first is to have knowledge of such great matters, for to this, according to the philosophers, is human nature naturally inclined....The second end, and the chief, is that it gives great aid in the knowledge of God the Creator....The third end of astrology and the least important is to ascertain certain dispositions of this lower and corruptible nature, whether present or to come, and nothing beyond that.... [Coopland's translation].⁴⁵

Oresme, like Christine, defines knowledge of God as the *summum bonum* of the natural philosopher. Both Wilkins and Godwin have made a complete separation of themselves from nature. Flying to the moon, an imaginative or ecstatic act in the world of Lucian, Macrobius, Dante and Christine de Pizan, in seventeenth-century writers becomes a possible material reality, the possibility of transporting physical bodies from point to point. In Godwin's day, schemes for flying to the moon were being seriously advanced by scientific thinkers. The 1640 edition of Wilkins's *The Discovery of a World in the Moone*, for example, contained an additional chapter outlining a device for such a lunar voyage.⁴⁶ Wilkins's 1648 work *Mathematical Magick: or, the Wonders that may be performed by Mechanical Geometry* also contains a description of a flying chariot.

Christine's journey, on the other hand, does not simply involve a movement from one physical locus to another. Christine's flight of speculation culminates in the knowledge of virtue: she undertakes an inward theological and spiritual journey. Stéphane Gompertz alerts us to the integrating and unifying function of Christine's epistemological journey in *Le Chemin de Long Estude*: 'the voyage [through the spheres] not only links the different regions of the universe, but also differing realms of knowledge; in this resides the journey's totalizing power'.⁴⁷ Elevated by her knowledge of the proper ethics of the prince, which she has heard debated in the Court of Reason, Christine undergoes both an inner and outer spiritualization in her travels to the heavenly spheres. Christine writes in the *Chemin de Long Estude*: 'through knowledge, the great treasure of the Understanding, better than gold, is engendered in our breasts; the fruit of such knowledge heals all wounds. Knowledge is the sun whose light illuminates the shadows of our thoughts with its fullness'.[48] The growth of inward knowledge involves building the bright, ethereal ladder of speculation in the mind's eye. Sebille suggests that Christine serve as a messenger to earth to report the debates in the court of Reason: Christine therefore truly becomes an 'influence', a spiritualized daemon or angel, transformed into the intermediate aetherial or pneumatic substance possessing the moral understanding that she will breathe into ('influer') the ears of earthly princes. Illuminated by ethical knowledge, she may also be likened to astrological rays transporting the heavenly influences to the terrestrial realm.

In Christine's conception of heaven, a strong link exists between divine knower and the known world; both are described as light and clear in her frequently employed Platonic light-as-knowledge metaphor. In the case of spiritual knowledge, the knower and the known are not distinct epistemological positions. Astronomy or astrology would not have to be studied in the heavens, since study implies a knowledge outside the individual; this separation could not occur in union with God. In heaven, the shining glorified bodies united to their souls in Christine's poem *La Prison de la Vie Humaine*, 'will find that they have all forms of knowledge, know all things perfectly, feel the infinite goodness of God'.[49]

In Christine's theory of knowledge, the long road of study requires moral and abstract thought, which then bridges the spatial and epistemological distance between the soul and heaven, the soul's final destination. As the two realms interrelate spiritually, there is no real concept of space or locus in Christine: distance between objects may be physical, but not spiritual, or vice versa. For Godwin and Wilkins, representing the triumph of the new physical, materialist astronomy in their celestial voyage narratives, the material universe is now distinct from the human soul.

References

1. Gaston de Paris, 'Chronique', *Romania* 10 (1881) 318: 'un ouvrage dont la valeur poétique est médiocre et qui n'a d'intérêt que pour l'histoire des idées et de l'instruction au XVe siècle'. Unless noted, all translations are my own.
2. Christine de Pizan, *Le Livre du Chemin de Long Estude*, ed. Robert Püschel (Paris 1887) 413-414: 'le feu et l'iave s'entreheent,/ A destruire l'un l'autre beent.'
3. See Edward Grant, *Planets, Stars and Orbs: The Medieval Cosmos, 1200-1687* (Cambridge 1994) 59: 'Medieval society's concept of the origin, structure, and operation of the world was drawn almost exclusively from the Aristotelian-Ptolemaic astronomical and cosmological tradition.' Bernardus Silvestris's *Cosmographia* has been translated by Winthrop Wetherbee, *The Cosmographia of Bernardus Silvestris: A Translation With Introduction and Notes* (New York 1973).
4. Richard Lemay, 'The True Place of Astrology in Medieval Science and Philosophy: Towards a Definition', *Astrology, Science and Society: Historical Essays*, ed. Patrick Curry (Woodbridge, Suffolk 1987) 57-78.
5. Both of these questions are intimately related to Aristotelianism. Aristotle had clearly stated the world was eternal and non-created, while the book of *Genesis* forced Christian theologians to argue for creation ex nihilo. Aristotle also denied the possibility of other worlds, arguing from his theory of the natural place of the elements. The heaviest element, earth, naturally falls towards the centre of the universe, i.e. the sphere of the earth, while light elements such as fire escape towards the heavens. Multiple worlds entailed multiple centers of attraction, in which case elements would be flying around the cosmos helter skelter. But denying the possibility of multiple created worlds placed a fundamental restriction on an omnipotent being. See Stephen J. Dick, *The Plurality of Worlds* (Cambridge 1982) and Pierre Duhem, *Medieval Cosmology: Theories of Infinity, Place, Time, Void, and the Plurality of Worlds*, ed. and trans. Roger Ariew (Chicago 1985) 431-510.
6. Marie-Josèphe Pinet, *Christine de Pizan 1364-1430: Étude Biographique et Littéraire* (Paris 1927) 423: 'Christine ne cesse de considérer Aristote comme le prince des philosophes.'
7. Christine de Pizan, *Christine's Vision*, trans. Glenda K. Mcleod, Garland Library of Medieval Literature, vol. 68, series B (New York 1993) 93 n.24.
8. *Chemin* 636-37. Jane Chance explores the question of female knowledge and its metaphors in 'Christine de Pizan as Literary Mother. Women's Authority and Subjectivity in "The Floure and the Leafe" and "The Assembly of Ladies"', *The City of Scholars: New Approaches to Christine de Pizan*, ed. Margarete Zimmermann and Dina De Rentiis (Berlin 1994) 245-259.
9. *Chemin* 1602-1610: 'Legiere estoit et portative/ Si qu'on la peust ertortillier/ Et porter sanz soy travaillier/ Par tout le monde, qui voulsist,/ Que ja n'empeschast ne nuisist,/ Non mie que de corde fust/ Ne d'autre file ne de fust;/ Ne je n'en congnois la matiere,/ Mais longue estoit, fort et legiere.'
10. Christine, following the practice of many medieval and Renaissance writers uses the terms 'astronomie' and 'astrologie' interchangeably. More precise writers followed Ptolemy's division in the *Tetrabiblos*, which circulated in the west in a widely read Latin translation entitled the *Quadripartitum*. Ptolemy clearly distinguished two cosmological sciences: the first part of the science of the stars consisted of the study of the appearance and movements of the celestial bodies with respect to the earth (what today we would call

astronomy); the second part considered the effects of the heavenly bodies on terrestrial events and human actions (what we would call astrology).
11. *Chemin* 1824-32: '[Sebille] tout me monstroit, et devisoit/ Des planetes les noms, la force,/ Et de moy enseignier s'efforce/ Les cours des estoilles mouvables/ Et des estans et des errables./ Si m'en dist les proprietez,/ L'effect, les contrarietez,/ Leurs forces et leurs influences/ Et leurs diverses ordenances.'
12. *Chemin* 1848-9: 'Car sience d'astrologie/ N'ay je pas a l'escole aprise.'
13. Charity Cannon Willard, 'Christine de Pizan: The Astrologer's Daughter' *in Mélanges à la Mémoire de Franco Simone*, vol. 1 (Genève 1980) 95-111; Charity Cannon Willard, *Christine de Pizan: Her Life and Works* (New York 1984) 17, 19-22, 97, 104; Edgar Laird, 'Astrology in the Court of Charles V of France, As Reflected in Oxford, St. John's College, MS 164' *Manuscripta* 34 (1990) 167-176.
14. Lynn Thorndike, *A History of Magic and Experimental Science*, 8 vols. (New York 1923 - 58) 3.585.
15. I am working from G.W. Coopland's paraphrase of Oresme's *De l'Espere* in *Nicole Oresme and the Astrologers: A Study of His Livre de Divinacions* (Liverpool 1952) 17-20.
16. *Chemin* 1934, 1943.
17. *Chemin* 1994. For an introduction to the history of this idea, along with excerpts from original texts, see Joscelyn Godwin's *Harmony of the Spheres: A Sourcebook of the Pythagorean Tradition in Music* (Rochester 1993).
18. *Chemin* 2010-11.
19. *Chemin* 2030-34.
20. *Chemin* 2095.
21. *Chemin* 2109-2118: '....aussi tost que l'omme naist/ Ou la femme, ja si grant n'est,/ Ceulx [les destinees] yci de sa vie ordenent/ Et sa droite fin lui assenent,/ Bonne ou male, selon les cours/ Ou les planetes ont leurs cours/A l'eure que l'enfant est né./ Mais toutefois Dieux, qui donne/ Leur a ce povoir, dessus est,/ Qui bien garde ce qui lui plaist.'
22. *Chemin* 2175-84.
23. For example, *Le Jugement du Roy de Behaigne* and *Le Jugement du Roy de Navarre*.
24. Barbara K. Altmann, 'Reopening the Case: Machaut's *Jugement* Poems as a Source in Christine de Pizan', *Reinterpreting Christine de Pizan*, ed. Earl Jeffrey Richards, Joan Williamson, Nadia Margolis, and Christine Reno (Athens 1992) p 137.
25. *The Man in the Moone: or a discourse of a Voyage thither by Domingo Gonsales The Speedy Messenger* (London 1638). All references to *The Man in the Moone* will be from the 2nd edition of 1657.
26. Lucian, *The True History*; Plutarch, *De Facie in Orbe Lunae*; Macrobius, *Somnium Scipionis;* Dante, *Divina Commedia*. Marjorie Hope Nicolson surveys some of this literature in *A World in the Moon: A Study of the Changing Attitude Toward the Moon in the Seventeenth and Eighteenth Centuries*, Smith College Studies in Modern Languages, vol. 17, no. 2 (Northampton, MA 1936).
27. *Dictionary of National Biography*, ed. Sir Leslie Stephen and Sir Sidney Lee, vol. 8 (Oxford 1917) 56 - 58.
28. *Planetary Astronomy from the Renaissance to the rise of Astrophysics, The General History of Astronomy*, ed. René Taton and Curtis Wilson, vol. 2, Part A: Tycho Brahe to Newton (Cambridge 1989) 5-7. See also J. L. E. Dryer, *A History of Astronomy from Thales to Kepler*, rev. W. H. Stahl, ed. 2 (New York 1953) 365-71 and Clarisse Doris Hellman, *The Comet of 1577: Its Place in the History of Astronomy* (New York 1971).

29. Godwin 68.
30. Godwin 65.
31. *Chemin* 1703-22.
32. See Thorndike 6.1-2, 83, 384.
33. H.W. Lawton, 'Bishop Godwin's *Man in the Moone*', *Review of English Studies* 7, no. 25 (1931) 41.
34. *Chemin* 1699-1700: 'comme une petite pelote,/ Aussi ronde q'une balote.'
35. Godwin 58-59.
36. Godwin 60.
37. *Kepler's Somnium: The Dream, or Posthumous Work on Lunar Astronomy*, trans. Edward Rosen (Madison 1967) 82 n.96.
38. *Dictionary of National Biography*, 21.264 -7.
39. Marjorie Hope Nicolson, *Voyages to the Moon* (New York 1948) 93.
40. John Wilkins, *The Discovery of a World in the Moone (1638), A Facsimile Reproduction with an Introduction by Barbara Shapiro* (Delmar 1973) 25.
41. Wilkins 54.
42. Wilkins 44.
43. Wilkins 40.
44. Christine de Pizan, *Le Livre de la Mutacion de Fortune*, ed. Suzanne Solente, t. 2 (Paris 1957) 7309-14: 'la superceleste/ Science...Ou est compris science entiere.' See also Christine de Pizan, *Le Livre des Fais et Bonnes Meurs du Sage Roy Charles V*, ed. Suzanne Solente, vol. 2 (Paris 1936) 18, 'comme l'entencion finale de sapience ou de methaphisique soit pervenir à cognoistre le gouvernement de la cause premiere, c'est Dieu le glorieux, la cognoissance de l'ordre des esperes celestes, auxquelles cognoiscences impossible est venir, senon après astrologie; et toutefois à astrologie nul ne puit parvenir s'ançois n'est philosophe, geomettre et arismetien; par quoy, comme il appert qu'en l'ordre des sciences astrologie et methaphisique sont tres haultes'.
45. Nicole Oresme, *Livre de Divinacions* in Coopland 113: 'la science du ciel a trois tres nobles fins. La premiere est avoir congnoissance de si tres belles choses car a ce est naturellement humain lignage enclin selon les philosophes.....La seconde fin et la plus principale d'astrologie est ce que elle donne grant aide a la congnoissance de Dieu le createur....La tierce fin d'astrologie et la moins principal est congnoistre aucunes disposicions de ceste basse nature corruptible presentes ou avenir et tant et nomplus....'.
46. Barbara Shapiro, 'Introduction,' Wilkins vi.
47. Stéphane Gompertz, 'La voyage allégorique chez Christine de Pisan', in *Voyage, Quete, Pelerinage Dans la Literature et la Civilization Medievales*, Senefiance No. 2, Cahiers du CUER MA (Paris 1976) 200: 'le voyage ne met pas seulement en contact les régions de l'univers mais aussi les domaines de la connaissance: c'est là que réside sa vertu totalisante.'
48. *Chemin* 5215-21: 'Par la quelle [la science]le grant tresor/ De conscience, meilleur que or,/ Est conceu en nostre courage,/ Dont le fruit tous maux assouage./ C'est le souleil par quel lumiere/ Ajourne o sa lueur pleniere/ Es tenebres de la pensee.'
49. *The Epistle of the Prison of Human Life* 60-61: 'ilz se trouveront avoir parfaite sapience, sachans toutes sciences, congnoissans toutes choses parfaitement, sentir l'infinie bonté de Dieu.'

Kepler's *Tertius Interveniens*

Ken Negus

Tertius Interveniens, written in 1610, is one of Kepler's most powerful and passionate treatises on astrology, written as a defence of the subject against extremists on both sides, on the one hand those who would condemn astrology altogether, and on the other those who accepted everything said and done in its name, no matter how preposterous. Hence he is the 'third party intervening', as indicated by the title.

The following extract is near the centre of the book, and is otherwise 'central' as an expression of the main tenets of Kepler's thought on astrology. In this short passage, he comments incisively on the following topics: the non-material, 'spiritual' nature of astrology; geometry as the all-embracing archetype through which the messages of the sky are communicated to earth; the horoscope as indelible 'imprint' on the soul of the infant being born; astrology as 'music'; the rationale of progressions and directions (symbolic measures used in prediction); astrological genetics; transits and mundane (political and historical) astrology.

Not to be forgotten when reading Kepler's arguments concerning astrology is his major role in the history of science. Although he is remembered primarily as an astronomer he also had much to say about biology, geology, meteorology, medicine and many other areas. His philosophical thinking also suggests much that is far ahead of his time, including prefigurations of Jungian psychology (archetypes and the collective unconscious). But let him, in slightly abridged form, speak for himself:[1]

Thesis 64. All powers coming down from above are ruled according to Aristotle's teaching: namely, that inside this lower world or earthly sphere there is a spiritual nature, capable of expression through geometry. This nature is enlivened by geometrical and harmonic connections with the celestial lights, out of an inner drive of the Creator, not guided by reason, and itself is stimulated and driven for the use of their powers. Whether all plants and animals, as well as the Earth's sphere, possess this faculty, I cannot say. It is not an unbelievable thing, for they have various faculties of this kind: in that the form in every plant knows how to put forth its adornment, gives the flower its colour, not materially, but formally, and also has a certain number of petals; nor [is it unbelievable] that the womb, and the seed that falls into it, has such a marvellous power to prepare all the body parts in appropriate form....The human being, however, with his soul and its lower powers has such an affinity with the heavens, as does the surface of the Earth, and this has been tested and proven in many ways, of which each is a noble pearl of astrology, and is not to be rejected along with [all of] astrology, but to be diligently preserved and interpreted.

Thesis 65. Above all, I might in truth flatter myself with having experienced this observation: that the human being, in the first igniting of his life, when he now lives independently for the first time, and can no longer live in the womb, receives the character and formation of the sky's whole stellar configuration, or the form of the conflux of radii on earth, and maintains it unto his grave. Afterwards this can be perceived in the formation of the face and the remaining bodily structure, as well as in the person's behaviour, habits and gestures, so that he might create, with his bodily form, corresponding attraction and charm for himself in the eyes of other people, and with his actions bring forth corresponding fortune. Then thereby (as well as from the mother's fantasies before the birth and from the rearing of the child thereafter), a great difference from other people is created, so that one person is brave, cheerful, joyful, self-confident; and another lethargic, lazy, neglectful, shy, forgetful, hesitant, and whatever other general characteristics there may be, which can be compared with configurations that are pleasant and exact, or complex and awkward, or also with the colours and motions of the planets. This character is not received into the body - for it is much too ungainly for that - but into the nature of the soul itself, which behaves like a point, so that she [the soul] might be transformed in points of the conflux of radii; and not only do these points impart reason to her, from which we human beings might be called reasonable above all other creatures, but she is to grasp in the first moment another kind of implanted reason - geometry - in the radii as well as in musical sounds [i.e. 'voices,' in the technical musical sense], without a lengthy learning process.

Thesis 66. Second, so it is with every plant, that it is on schedule when it is to ripen or blossom. This time is prescribed to it at its creation, and by external warmth and other means it is lengthened or shortened, but can never be totally altered. In like manner the human being's nature, upon entry into life, receives not only an instantaneous image of the sky, but also its movement, as it appears down here on earth for several days in a row, and in certain years derives from this moment the manner of outpouring this or that humour; these years are precisely and sharply indicated, based on the projection of the first few days [of life]. This is a truly marvellous thing, and is like an image or outflowing of the natural proportion of a day to a year. Thus this short time or 'tempus typicum' in human nature with all its parts is multiplied by 365; and all of natural life, out of this multiplication, remaining rigidly in its memory, is deducted and unwound as from a ball of yarn, so that then the whole future life, insofar as it deals with natural matters, in the course of a quarter-year is wound up and stored in a little bundle. Such a causality and natural proportion cannot, however, be applied to the profections,[2] for not the Ascendant and not the Sun, but Jupiter makes its revolution in 12 years, as the Moon does in 28 days, and accordingly the best of the profections should be assigned to the transits; the rest is useless noise. I have often harboured the thoughts that there is nothing to directions because we must reach out so far for their causality, and one cannot accommodate them any differently. But I must confess that nonetheless the causality resembles nature, because it requires a natural proportion; and that our experience is so clear, that they are not to be denied as true for the astrologers.

Thesis 67. Third: this is a curious thing that the nature [of the human being], which receives this character [of the sky], also favours its relatives by some similarities in the celestial constellations. When the mother is great with child and her time has come, then nature seeks out a day and hour for the birth that is comparable celestially with that of the mother's father or her brother.

Thesis 68. Fourth, every [human] nature knows not only its celestial character, but also every day's configurations and motions in the sky so well, that as often as a transiting planet comes into its character's Ascendant or other prominent place, especially into radical points, it [the nature] accepts it and is thereby variously affected and stimulated.

Thesis 69. Fifth, there is also the experience that every strong conjunction, by itself, without considering the relationship to a particular person, stimulates people in general (where a nation lives together in an ordered society), and makes them capable of acting as a community that is unified just as the stars then shine together. This was discussed in detail in my book *De stella serpentarii*. Thus I have seen many examples of epidemics in which the humours are stirred up more when strong stellar configurations are present (that is to say, the human natures are stimulated to drive out the humours). In like manner all these points - and many more could be cited, being from the same cause - and the possibility of one coming from the other could be proven and defended.

References

1. Translation by Ken Negus, 1997.
2. A forecasting technique in which the houses of the horoscope represent the years of life, beginning with the first house and the first year of life.

Belief in Astrology:
a social-psychological analysis

Martin Bauer[1] and John Durant[2]

Abstract

Social scientists have suggested several different hypotheses to account for the prevalence of belief in astrology among certain sections of the public in modern times. It has been proposed: (1) that as an elaborate and systematic belief system, astrology is attractive to people with intermediate levels of scientific knowledge [the superficial knowledge hypothesis]; (2) that belief in astrology reflects a kind of 'metaphysical unrest' that is to be found amongst those with a religious orientation but little or no integration into the structures of organized religion, perhaps as a result of 'social disintegration' consequent upon the collapse of community or upon social mobility [the metaphysical unrest hypothesis]; and (3) that belief in astrology is prevalent amongst those with an 'authoritarian character' [authoritarian personality hypothesis].

The paper tests these hypotheses against the results of British survey data from 1988. The evidence appears to support variants of hypotheses (1) and (2), but not hypothesis (3). It is proposed that serious interest or involvement in astrology is not primarily the result of a lack of scientific knowledge or understanding; rather, it is a compensatory activity with considerable attractions to segments of the population whose social world is labile or transitional; belief in astrology may be an indicator of the disintegration of community and its concomitant uncertainties and anxieties. Paradoxical as it may appear, astrology may be part and parcel of late modernity.

1. Introduction

Across the industrialized world, astrology has attractions for large numbers of people. Horoscopes are read by millions; astrologers are personally

---

[1] London School of Economics, Department of Social Psychology

[2] Science Museum and Imperial College, London

consulted by tens or hundreds of thousands; rumour has it that the London City is a booming place for astrological consultancy; even the wives of Presidents[1], it appears, may consult with astrologers before advising their husbands on how to conduct affairs of state. In all these situations astrology seems to offer a degree of certainty where uncertainty prevails. To many scientists and science educators who are concerned about the public understanding of science, the enduring popularity of astrology is an affront. How can it be, they ask, that in the last decade of the 20th century so many people are still prepared to embrace pre-scientific and even frankly superstitious belief systems?

Faced with the task of accounting for the enduring popularity of astrology, it is tempting to invoke the phenomenon of 'anti-science' - that is, active resistance to the principles and practises of science. In this context, it may be significant that the first of a series of US-Soviet conferences on the social and political dimensions of science and technology, which was held at the Massachusetts Institute of Technology in May 1991, was devoted to 'Anti-Science Trends in the United States and the Soviet Union'. Significantly, the two parallel keynote addresses to this conference - by Gerald Holton, of Harvard University, and Sergei Kapitza, of the Institute for Physical Problems (Moscow) - pointed to the need for a critical understanding of the phenomenon of anti-science. According to Holton anti-science in the US is symptomatic of a long-standing struggle over the legitimacy of the authority of conventional science;[2] while for Kapitza, anti-science in the east is part and parcel of the wider social and political transformation of the former Soviet Union.[3]

In a recent BBC radio programme prominent representatives of churches, science, and the arts discussed the apparent popularity of astrology and parasciences in Britain under the label 'pre-millennium tension' [PMT].[4] Ironically, on the issue of astrology and parasciences, the traditionally polarised positions of science and religion converged. It seems that present day astrology claims the territory which makes the Church and Science equally nervous. Albeit, the nervousness may have different sources.

In this paper, we investigate the phenomenon of popular belief in astrology in Britain in the late 20th century. Our evidence concerning the place of astrology in British culture is derived from the results of a 1988 national random sample survey designed to estimate levels of public interest in, understanding of and attitudes towards science and technology. In the course of this survey several questions were asked about astrology[5]. The results of these items enable us to explore three different sociological hypotheses which have been advanced to account for the prevalence of belief in astrology amongst certain sections of the public: first, that astrology is attractive to people with intermediate levels of scientific knowledge [superficial knowledge hypothesis]; second, that astrology is attractive to people who possess what has been termed 'metaphysical unrest' without integration into a Church; their unrest could therefore be considered free-floating [metaphysical unrest hypothesis]; and third, that belief in astrology is prevalent amongst people with authoritarian personality characteristics [authoritarian personality hypothesis].

Astrology must be considered the "grandmother" of modern science in at least two aspects: its concern with regularities in the universe, and its attempt to deal with these regularities numerically. Keith Thomas observed that 'at the beginning of the 16th century astrological doctrines were part of the educated man's picture of the universe and its workings'; London was a booming centre of astrological divinations for a mainly elite clientele of Court, nobility and Church until its decline in the mid-17th century.[6] In one sense it is not surprising that in a country that prides itself on tradition and continuity we find residuals or even revivals of such activities in the late 20th century. In this paper we try to locate contemporary belief in astrology in order to understand its social and psychological functions; while temporarily abstaining from evaluations of the belief itself.

We begin by defining our measures of public belief in astrology, and then proceed to use these measures to explore the three hypotheses.

2. Measuring Popular Belief in Astrology

The British survey was conducted in the early summer of 1988. The sample of 2009 respondents was designed to be representative of the adult population of Britain over the age of eighteen. The survey was conducted by means of face-to-face interviews lasting between forty minutes and one hour. The questionnaire covered a wide variety of topics in the general field of science and technology. In particular, it developed a multi-item scalar measure of scientific understanding. Further details of the survey methodology and the results on public understanding of science have been published elsewhere.[7,8,9,10]

So far as the present study is concerned, the following items from this national survey are of particular interest. First, respondents were asked 'Do you sometimes read a horoscope or a personal astrology report?'. Those who responded positively were then asked (a) how often they read a horoscope or personal astrology report [frequency] and (b) how seriously they took what these reports said [seriousness]. 73% of respondents claimed to read a horoscope or personal astrology report. 21% said that they would read it 'often', 23% 'fairly often', 29% 'not often', and 27% did not read it 'at all'. Hence, 44% claimed to do so often or fairly often. However, a rather smaller number of respondents (6%) claimed to take what horoscopes or personal astrology reports said either 'seriously' or 'fairly seriously'. 67% took it not very seriously, and 27% took it not at all seriously. This result points immediately to the problematic status of astrology in the minds of many of those who take at least some personal interest in it.

Figure 1: the combined percentages of respondents for two questions: 'how frequently do you read astrology columns?' and 'how seriously do you take it?'

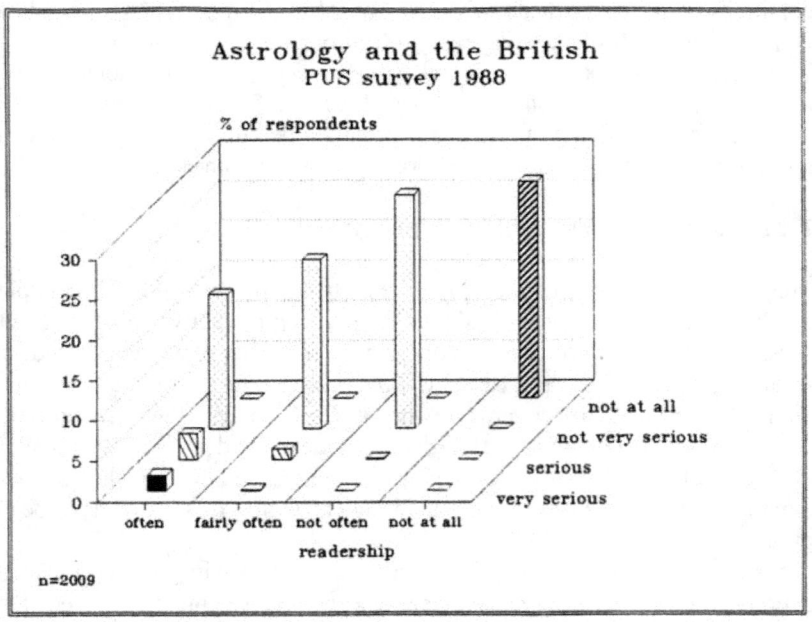

In order to accommodate these results in a useful way, we have combined them into a single scalar measure. Figure 1 brings together the results on readership and seriousness, which we combined into a 5 point-scale of belief in astrology. The scale is derived from the readership and seriousness results in the following way: those who reported that they read horoscopes often or fairly often and that they took them seriously or fairly seriously are ranked 5 (serious believers 5%); those who reported that they read horoscopes often and that they took them not very seriously are ranked 4 (non-serious believers, 18%); those who reported that they read horoscopes fairly often and that they took them not very seriously are ranked 3 (non-serious believers, 21%); those who reported that they read horoscopes not very often and that they took them not very seriously are ranked 2 (non-serious believers, 29%); and those who reported that they did not read horoscopes at all are ranked 1 (non-believers, 27%).[11] With around 5% of the population or 2.5-3 million, the constituency of serious believers in astrology is a small minority compared to the constituency adhering to basic religious creeds such as 'God', a 'life after death' or 'miracles', which includes half or more of the British population.[12] For much of the following analysis the 5-point scale is reduced by pooling 1+2, 3, and 4+5 into a 3-point scale.

Another item in the survey invited respondents to estimate the scientific status of astrology (which was defined as 'the study of horoscopes') on a 5 point-scale, from 'not at all scientific' to 'very scientific'. 32% of respondents stated that astrology was not at all scientific (scale point 1), while 13% stated that it was very scientific (scale point 5); 18% said it was in between (scale point 3); a further 17% tended towards 'not scientific' (scale point 2), and 14% tended towards 'scientific' (scale point 4); 5% did not know.

Our survey incorporated two standard measures concerning religious belief and religious integration. Religious belief was constructed as a scalar measure on the basis of responses to the following agree/disagree items: 'spiritual values taught by religion are important'; 'there is no such thing as a God'; 'people should rely more on the power of prayer'; and 'Adam and Eve never really existed'.[13] Religious affiliation was constructed as a scalar measure on the basis on the following items: 'Do you regard yourself as belonging to any particular religion?'; and (if yes), 'Apart from such special occasions as weddings, funerals and baptisms, how often nowadays do you attend services or meetings connected with your religion?'.[14]

Finally, the survey comprised two standard scales on 'authoritarianism-egalitarianism' and 'social efficacy'. Authoritarianism is indicated by consistently agreeing with statements such as 'censorship of film and magazines is necessary to uphold morality' or 'school should always teach children to obey authority'. Social efficacy is indicated by disagreeing with statements such as 'I feel it's very difficult to have any real influence on what other people do or think' or agreeing with 'people like me can influence the government by taking an active part in politics'.

3. Exploring the Basis of Popular Belief in Astrology

Equipped with the measures that have been described above, we can begin to explore the basis of popular belief in astrology. We shall do this by considering in turn three different hypotheses that have been advanced to account for this phenomenon.

i.   **Superficial Knowledge**

It has been claimed that belief in astrology is the product of a relatively slight or superficial acquaintance with the stock of modern scientific knowledge. On this view, people with what might be termed an intermediate level of scientific understanding may be attracted by astrology because it possesses many of the 'trappings' of orthodox science (systematic structure, predictive power, numeracy etc.); but they may be insufficiently well equipped to see that these things really are the 'trappings' rather than the substance of genuine science. Thus, in his classic paper of 1957 on the *Los Angeles Times* Astrology Column as an example of 'secondary superstition', Theodor Adorno wrote as follows:

> While the naive persons who take more or less for granted what happens hardly ask the questions astrology pretends to answer and while really educated and intellectually fully developed persons would look through the fallacy of astrology, it is an ideal stimulus for those who have started to reflect, who are dissatisfied with the veneer of mere existence and who are looking for a 'key', but who are at the same time incapable of the sustained intellectual effort required by theoretical insight and also lack the critical training without which it would be utterly futile to attempt to understand what is happening.[15]

We may pass over what seem by today's standards the somewhat elitist and patronising tones of Adorno's analysis. What concerns us here is whether the basic prediction - that astrology is attractive to people with intermediate levels of scientific understanding - is fulfilled. If that were the case, we would expect belief in astrology to be positively correlated with knowledge of science up to a certain level of scientific knowledge, beyond which this correlation becomes negative. In other words, we would expect a non-linear inverted U-shape relationship shown between scientific knowledge and the status of astrology.

This issue may be addressed by comparing the results of our question on the scientific status of astrology with the results of our multi-item scalar measure of scientific understanding. Figure 2 shows these results, compared with those for a similar item on the scientific status of physics. While there is a linear relationship between scientific understanding and the perceived scientific status of physics, there is a curvilinear relationship between scientific understanding and the perceived scientific status of astrology. In other words, our data do indeed bear out Adorno's hypothesis.

It should be noted that Figure 2 gives the proportions of respondents who ranked astrology and physics as 'very scientific'. We can learn a little more by comparing these results with those for other available options concerning the scientific status of astrology. Figure 3 shows the results for three groups of respondents: those who stated that astrology is not scientific (responses 1 + 2); those who stated that astrology is neither scientific nor unscientific, or who said they didn't know (neither + don't know); and those who stated that astrology is scientific (responses 4 + 5). Those with low levels of understanding have a strong tendency to avoid a definite judgement about astrology; while those with high levels of understanding have a strong tendency to state that astrology is unscientific. Amongst those with intermediate levels of understanding, there is less obvious consensus: some think astrology is scientific, some think it is not, and some don't know.

So much for the perceived scientific status of astrology. What, we may ask, about belief? Figure 4 compares belief in astrology with scientific understanding measured by a 28-item knowledge quiz.[16] As we might expect overall there is a negative correlation between scientific understanding and belief in astrology ($r = -.21$). However, on closer inspection it emerges that this negative correlation applies only to the

Figure 2: the scientific status attributed to physics and astrology in relation to the level of understanding of science; percentage of respondents saying 'scientific' or 'very scientific' combined.

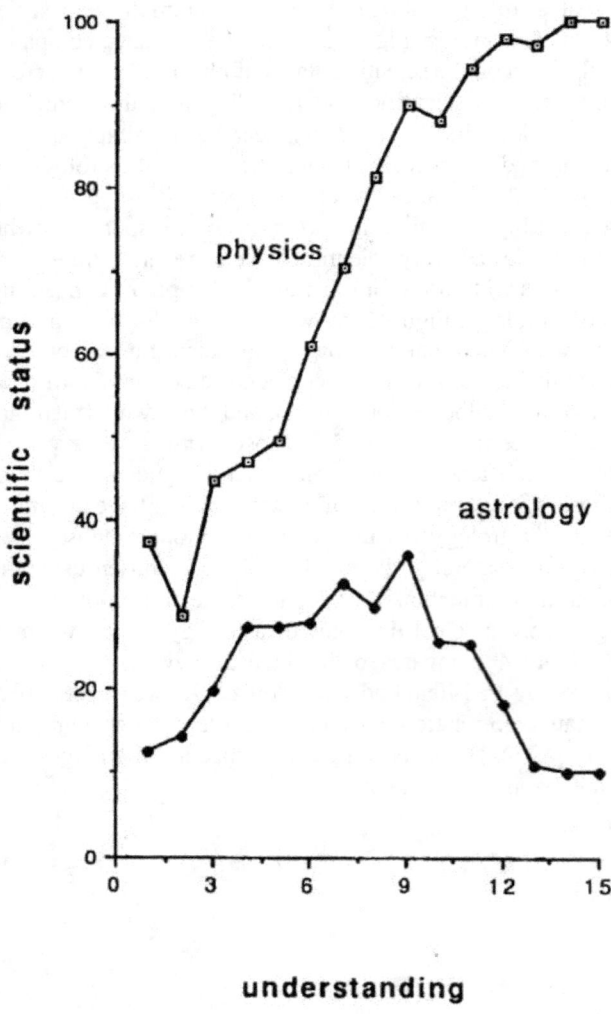

Figure 3 shows the percentage of respondents saying that 'astrology is not scientific', 'don't know' or 'astrology is scientific' in relation to levels of understanding of science.

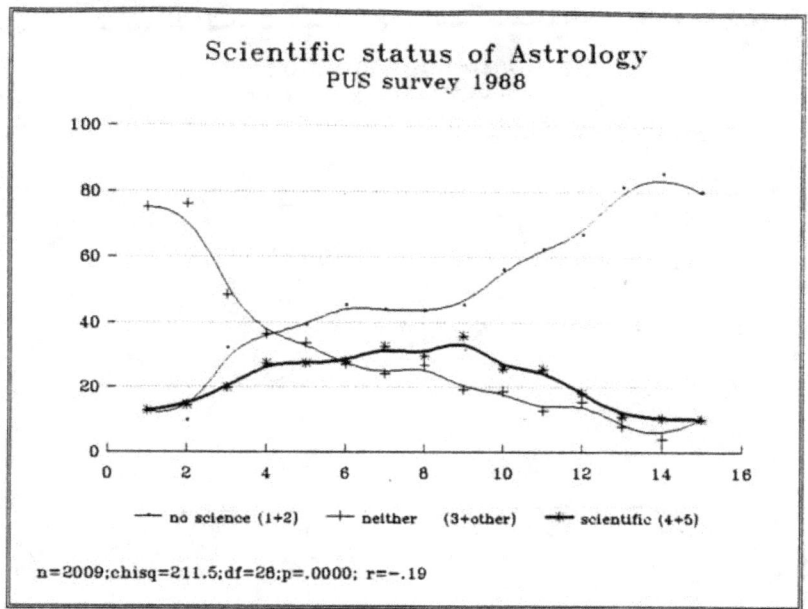

upper half of the understanding scale. We may wish to ignore the sudden jump of belief in astrology at the very top of the knowledge scale, which is based on a too few observations to be significant. However, within the 50% of the general public whose relative scientific understanding is below average, there is no correlation at all between levels of understanding and belief in astrology. This is a pointer to a potential problem with measures of scientific literacy that incorporate questions on the scientific status of astrology.[17] Empirically, astrology and science are not mutually incompatible at least at lower levels of scientific enculturation. To use astrology as a threshold measure for 'scientific literacy' may be justifiable on normative grounds, but it ignores the social phenomenon of compatibility or incompatibility between these two forms of knowledge, which is itself a significant cultural variable. We expect the correlation to differ across cultural contexts.[18]

Figure 4 shows the average intensity of belief in astrology in relation to the level of scientific understanding.

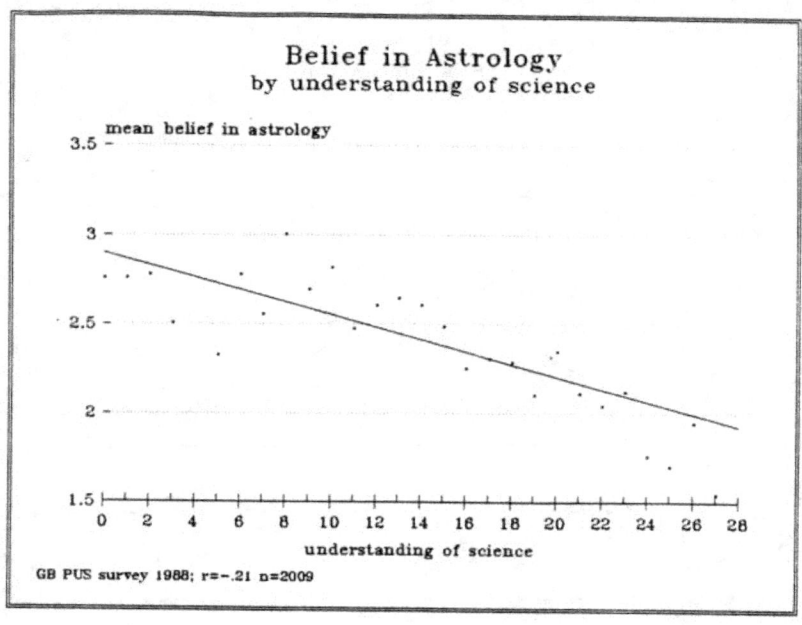

ii. 'Metaphysical Unrest'

It has been claimed that astrology has particular attractions for people who are alive to religion but who are poorly integrated into the institutional structures of a religious community. In this category are, for example, those who have been brought up in a particular religion and retain a religious outlook on life, but who for one reason or another (including social mobility or the collapse of community) have ceased to be closely tied to the particular church in which they were raised. Thus, Maitre and Boy & Michelat have observed in France of the 1960s and 1980s and Schmidtchen in Germany of the 1950s that astrology tends to be less popular amongst those who are closely integrated into the institutions of organized religion. The French characterize astrology as a petit-bourgeois phenomenon of social uncertainty, social isolation and individualisation.[19] According to Valadier, this result is consistent with the hypothesis that astrology feeds upon a free-floating 'metaphysical unrest', or a desire to recover a sense of the sacred and a sense of unity on the part of people whose life world no longer provides for these experiences; Pollack sees it as one among many forms of religiosity-outside-the-church in the context of the collapse of old certainties in Eastern Germany.[20] Based on these previous observations, we would expect to find serious inclinations towards astrology most prevalent among religious believers with little or no religious integration.

We may put this hypothesis to the test in the context of our British data. Our data show that there is a very slight tendency for belief in astrology to be greater amongst those with higher levels of religious belief (r = 0.10). However, inclination towards astrology is highest amongst those with intermediate levels of integration into the institutions of organized religion. Putting these results together, Figure 5 shows average belief in astrology in relation to both religious belief (1 = low; 3 = high) and religious integration (1 = low; 3 = high).

Figure 5: the average intensity of belief in astrology in relation to religious belief and religious integration

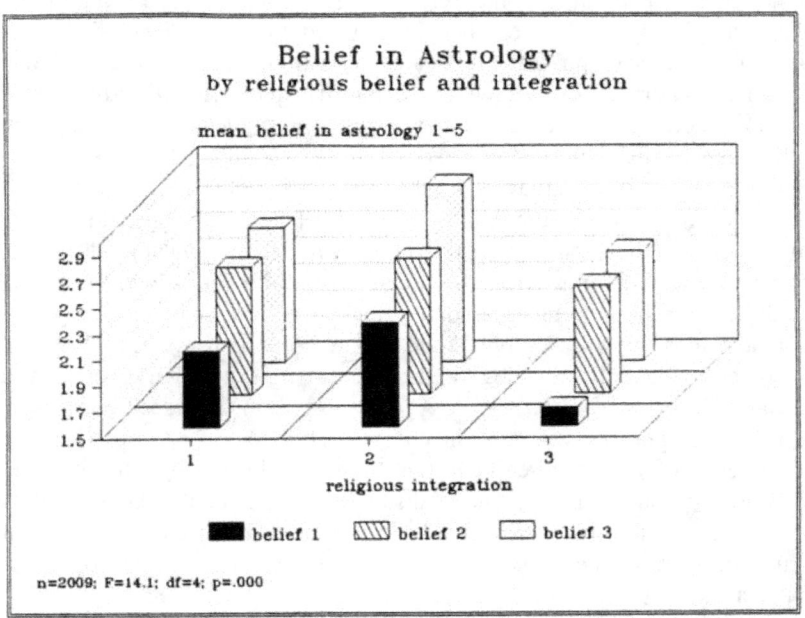

We see that belief in astrology is highest amongst those who combine strong religious belief and intermediate or low religious integration. The fact that an intermediary level of integration is associated with highest level of belief in astrology is perhaps unexpected. On the other hand, it may be that having one foot in the Church and the other outside it may be the very situation of social uncertainty which Valadier takes as diagnostic for present-day belief in astrology. To this extent, therefore, we are able to confirm Valadier's hypothesis and Schmidtchen, Maitre and Boy & Michelat's results suggesting that astrology has particular attractions for those who may be experiencing free-floating metaphysical unrest. Needless to say, our data do not permit us to explore the sources of such unrest in the lives of our respondents. This is an area where qualitative and biographical research may be more revealing.

iii. The Authoritarian Personality

The third and last hypothesis that we shall consider takes us back to the work of Theodor W Adorno. In the course of his analysis of astrology, Adorno noted that in general terms the astrological ideology resembles, in all its major characteristics, the mentality of the 'high scorers' of The Authoritarian Personality'. In addition to what he believed to be the narcissism, self-absorption, naive empiricism and fatalism of astrology, Adorno pointed to its tendency to attribute everything negative in life to external, mostly physical circumstances. In these and other ways, he suggested, astrology had affinities with the authoritarian personality.[21]

Once again, our data may be used to test this hypothesis since the survey contained a standard battery of psychological items designed to provide measures of authoritarianism-egalitarianism and 'social efficacy', defined as personal sense of control over the social world. The data shows that in our study there is no significant tendency for belief in astrology to be greater amongst those who score higher on the authoritarianism scale. We find, however, that belief in astrology is stronger amongst those who score low on social efficacy ($r = -.21$). Astrology, it would seem, is indeed particularly attractive to persons with certain characteristics, namely those who have little sense of control over their lives. Thus, Adorno's hypothesis is not supported by our data, while the fatalism element was confirmed. Given that this famous authoritarian personality syndrome is more complex than our crude measure suggests, we suggest that further work is needed on this subject.

4. Characterizing the Believers in Astrology

According to our results, the field in which the believers in astrology are generally to be found is one in which people possess intermediate levels of scientific understanding, high levels of religiosity, and low levels of religious integration. But what sorts of people are actually to be found within this field? In addition to what has already been said about personality, our data suggest that women are more likely to believe in astrology than men. Among the believers in astrology [scale 4+5] 77% are women; among the declared sceptics 73% are men [scale 1+2]. With the exception of clerks (a high proportion of whom are, of course, women) self-employed, skilled and semi-skilled workers are in that order more likely to believe in astrology than people in professional and managerial occupations. It is interesting to note here that according to Boy & Michelat, different social strata are associated with different sorts of 'para-interests': in France astrology is the pursuit of the less educated, while para-science is the pursuit of the highly educated. Our data do not allow us to compare this result with the situation in Britain.

These simple correlations are difficult to interpret because of the notorious problem of confounding variables. In other words, it may be that we find a correlation between belief in astrology and social class only because both in turn are related to some third factor (such as education, or social efficacy). To reduce the ambiguity of our results, we have subjected our data on belief in astrology to a form of statistical analysis (Logit modelling) which is designed to analyse differences between two unequally distributed groups.[22] In this case, we wish to analyse the contributions to differences in astrological belief of each of a series of independent variables. Each independent variable is assessed individually, whilst possible effects from all other variables are controlled. This analysis ranks independent variables in order of importance, and it excludes variables which are found to make no statistically significant contributions.

We used a Logit model in which differences between sub-sets of the sample with respect to belief in astrology were analysed with the following independent variables: interest in science; understanding of science; religious belief; religious integration; authoritarianism; social efficacy; age; gender; marital status; social class; educational level; and nature of work (i.e. full/part-time). Comparing the extreme groups of serious astrology believers (ranked 5) with non-believers (ranked 1 + 2) in this way, we obtain the following results. The variables which are relevant for the model are in order of importance: (1) gender, (2) religious belief, (3) living alone or in partnership, (4) age, (5) religious integration, and (6) the attributed scientific status of astrology. All other variables are irrelevant in explaining the difference between serious believers and sceptics. Note that the religious variables remain important, while personality and scientific understanding fall out of the equation. This indicates that the 'metaphysical unrest' hypothesis may be the strongest of the three hypotheses.

Comparing the category of playful, non-serious believers in astrology (ranked 4) with the sceptics (ranked 1 + 2), we obtain slightly different results. Again in order of importance the following variables are relevant: (1) gender; (2) marital status; (3) social efficacy; (4) educational level; and (5) attributed scientific status of astrology. In distinguishing between the playful and curious approach to astrology and the sceptics we lose the religious variable from the equation and gain education and efficacy.

At least as significant as the list of items that appear in these analyses is the list of items that do not. From these results, it would appear that interest in science and scientific understanding are not significant contributors to variations in belief in astrology. This, in turn, casts serious doubt on the advisability of employing measures of belief in astrology as constituent items in larger constructs concerned with scientific understanding or scientific literacy.

On the basis of these results, we can risk a caricature of believers in astrology. Serious believers in astrology tend to be: female rather than male; single rather than living with partners; younger rather than older; and religiously motivated rather than indifferent; and inclined to attribute scientific status to astrology. The non-serious and playful astrology consumer also tends to be female and to live alone, to be less educated, less in control of their affairs, and to consider astrology to be more scientific than the sceptics allow.

5. Conclusion

We began by citing recent concerns at the rise of astrology as an anti-science phenomenon, East and West. Kapitza suggests that in part the rise of anti-science in the (former) Soviet Union may be explicable in terms of the ideological collapse of the Soviet empire. Such a collapse may be expected to have left an intellectual and spiritual vacuum, and this in turn will have helped to bring about a certain amount of social disintegration. Similarly, Holton proposes that the anti-science phenomenon in the United States should be understood as part of a deeper opposition both to the authority of science and to a certain conception of modernity. Both of these analyses invite us to consider popular belief in astrology as a great deal more than the passive result of mere ignorance.

In general, we suggest that there are three different ways of approaching the problem of popular belief in astrology. First, it may be regarded positivistically, as an anachronistic survival of a pre-scientific world-view. In this context, popular belief in astrology is seen as an atavistic phenomenon. Second, it may be regarded anthropologically, as an alternative world-view deserving of attention and respect in its own right. In this context, we are required to make no value-judgements about the respective merits of non-scientific and scientific positions. Third, it may be regarded sociologically, as one among a number of potential compensatory activity that may be attractive to individuals who are struggling to come to terms with the uncertainties of life in late modernity.

In this paper, we have inclined towards the last of these approaches. Belief in astrology is rather a matter of the moral fabric of modern society than of scientific literacy. It seems that in Britain, as in Germany or France, belief in astrology is prevalent among particular social groups; groups which, as we have indicated, may be experiencing difficulty in accommodating their religious feelings to life in an uncertain post-industrial culture. Paradoxical as it may seem, therefore, we conclude that popular belief in astrology may be part and parcel of late modernity itself.

References

1. An earlier version of this paper was given to the Annual Meeting of the American Association for the Advancement of Science in Chicago, 9 February 1992; at the time it was common currency that Nancy Reagan, the wife of former President Reagan, was consulting with astrologers on matters of US state affairs.

3. Kapitza S (1991) 'Anti-science trends in the USSR', *Scientific American*, 265, 2, August, 18-24.

4. 'Moral Maze', 14 November 1996, BBC4, 9.00-10.00; moderated by Melvin Bragg. This term is a slightly sexist pun on the medically controversial 'pre-menstrual tension'.

5. Acknowledgement: The 1988 British national survey of public understanding of science is a joint Science Museum/University of Oxford and Community Planning Research (SCPR) project funded by the Economic and Social Research Council, grant numbers: A 09250013 and A 418254007.

6. Thomas K (1971) *Religion and the Decline of Magic. Studies in popular Beliefs in the 16th and 17th Century*, London, Penguin, 337ff.

7. Durant J R, Evans G A and Thomas G P (1989) 'The public understanding of science', *Nature*, 340, 11-14.

8. Evans G A & J R Durant (1989) *Understanding of Science in Britain and the USA, British Social Attitudes: Special international report,* edited by R Jowell et al., Aldershot, Gower, 105-119.

9. Durant J R , G A Evans and Thomas G P (1992) 'Public Understanding of science in Britain: the role of medicine in the popular representation of science', *Public Understanding of Science*, 1, 2, 161-183.

10. Evans G and J Durant (1995) 'The relationship between knowledge and attitudes in the public understanding of science in Britain', *Public Understanding of Science*, 4, 1, 57-74.

11. To measure the internal consistency of the 'belief in astrology' scale we use Cronbach Alpha's Reliability Coefficient: 0.92. Alpha is a measure for the covariance among all the items in the scale; Cronbach L J (1951) Coefficient alpha and the internal consistency of tests, *Psychmetrica*, 16, 297-334.

12. Greeley A (1992) 'Religion in Britain', Ireland and the USA, in: R Jowell, L Brook, B Prior, B Taylor (eds) *British Social Attitudes*, the 9th report, Altershot, Dartmouth, 51-70.

13. Internal consistency of the religious belief measure: Cronbach Alpha's reliability coefficient = 0.70.

14. Internal consistency of religious integration: Cronbach Alpha reliability coefficient = 0.73.

15. Adorno T W (1957) 'The stars down to earth, the Los Angeles Times astrology column, a study in secondary superstition', *Jahrbuch fuer Amerikastudien*, Heidelberg, 2, 19-88, reprinted in R.Adorno, *The stars down to earthand other essays on the irrational in culture,* edited Stephen Crook (Routledge, London and New York, 1994).

16. See Durant, Evans, and Thomas (1989) op.cit.

17. Miller J D (1983) 'Scientific literacy: a conceptual and empirical review', *Daedalus*, 112, 3, 29-48; Miller J D (1991) 'The public understanding of science and technology in the United States', 1990, A report to the National Science Foundation, Dekalb, Illinois, February 1991.

18. We do recall from the Chicago meeting in 1992 that in the discussion an Indian theoretical physicist was quite irritated and outspoken about the tacit assumption in much of the discussion according to which science and astrology were incompatible. He made reference to the Indian context where Brahmanic knowledge traditions seem to have no problem of compatibility between modern science and astrology.

19. Maitre J (1966) 'La consommation d'astrologie dans la societe contemporaine', *Diogenes*, 53, 92-109; Boy D & G Michelat (1986) 'Croyance aux parasciences: dimensions sociales et culturelles', *Revue Francaise de Sociologie*, 27, 2, 175-204; Schmidtchen G (1957) 'Soziologisches ueber die Astrologie', *Zeitschrift fuer Parapsychologie und Grenzgebiete der Psychologie*, 1, 47-72.

20. Valadier P (1987) *L'eglise en proces. Catholicism et societe modern*, Paris, Flammarion; Pollack D (1990) 'Vom Tischrucken zur Psychodynamik. Formen ausserkirchlicher Religiositaet in Deutschland', *Schweizerische Zeitschrift fuer Soziologie*, 1, 107-134.

21. Adorno T W , Frenkel-Brunswik E, Levinson D J and Sanford R N (1950*) The Authoritarian Personality*, New York, Harper.

22. Aldrich J H and F D Nelson (1989) *Linear probability, logit and probit models*, New Bury Park, Sage.

Visions of the Future: Almanacs, Time, and Cultural Change, **Maureen Perkins**, Oxford: Clarendon Press, 1996, 270 pp.; hardback; £40; ISBN: 0-19-812178-4

This is a brilliant book, combining thorough scholarship with original insight. It should deepen our understanding of a remarkable number of subjects. Essentially, it concerns a key part of the process of rationalisation that has been so instrumental in producing what we now recognize as modernity.

Perkins has much to tell us about astrological almanacs, to whose importance Keith Thomas first alerted us; in this capacity, she builds on and extends the excellent work of Bernard Capp. There is also fascinating material here on comic almanacs and Australian almanacs - the latter including an example of cultural influence by a colony (in the person of James Ross) on metropolitan discourse.

But more important is her use of almanacs to gain access to the world of popular belief, and the tensions in its relationships with elite opinion. Here the pioneer was Peter Burke's *Popular Culture in Early Modern Europe* and Perkins' book easily holds its own with subsequent scholarship by David Vincent and others. (I can't help feeling it a pity, however, that she passed over E. P. Thompson's apt refinement of 'popular' as 'plebeian'.)

At the heart of her account is the campaign by the Society for the Diffusion of Useful Knowledge in the third decade of the nineteenth century against the Stationers' Company, monopolist publisher of almanacs. Led by Lord Brougham and Charles Knight, the SDUK targeted such long-standing annual titles as *Poor Robin, Partridge's* and especially *Vox Stellarum*, popularly known as *Moore's*, with its mysterious hieroglyphic and astrological prophecy. In 1800, a minimum of one person in every seven in England bought an almanac - which was read, of course, by several more - and far and away the most popular was *Moore's*. In 1838, its best year, it sold over half a million copies, netting the Company of Stationers £6,414.

Significantly, the editorial voice of *Moore's* was unimpeachably Whig, comprising a set of convictions shared by the SDUK. But the latter had correctly identified the former as a major site and source of 'the superstitions of the vulgar' (in the characteristic terms of the puts it, was nothing less than 'a transformation of consciousness, from one which was connected to a pre-Enlightenment world of correspondences and humours perpetuated by popular almanacs, to one in which empirical observation and rational enquiry were the standard....[and] in which the natural world could be placed without recourse to 'irrational' concepts.' (p. 58)

Of course, the SDUK's empirico-rationalism was far from neutral, proceeding by a series of conflations linking 'useful', 'rational', 'scientific' and 'real'. In other words, this was a hegemonic struggle to replace one particular social construal of reality with another. (I should add, however, that Perkins is no wild-eyed student of cultural studies, however; a more sober and thoroughly documented account would be hard to imagine.) In this context, the dividing line between rationality and 'superstition' was bitterly contested. In a fascinating chapter on weather, Perkins tells the sad story of Admiral Robert Fitzroy, who pioneered efforts to take its prediction out of the hands of countrymen, astrologers and amateurs. In 1865, harried mercilessly by the press as a covert weather-prophet (and by astrologers on his other flank), Fitzroy took his own life.

Predictably, the overall results of the SDUK campaign were uneven and complex. In 1872, *Moore's* finally dropped the astrology, only to be severely punished by readers: sales dropped steadily to only 50,000 in 1895. It was farmed out to another publisher in the early twentieth century who re-introduced 'the voice of the stars', and still appears annually, though with nothing like its former circulation or influence. Meanwhile, in the 1830s, judicial astrology re-appeared in the metropolitan heartland, courtesy the new almanacs of Zadkiel and Raphael. These had a middle-class readership, and Perkins underestimates the significance of their success, which astounded Charles Knight; she could have made more use of them in grasping the complexity of mid-nineteenth-century middle-class discourse. She also succumbs to the temptation (which seems nigh-well irresistible to historians in this field) to perceive the 'death of popular astrology', this time in 1869-70 (p. 119); the evidence to the contrary in every daily tabloid newspaper, and even some broadsheets (to the disgust of others). True, Sun-sign columns aren't precisely early modern moon- and star-

Overall, however, Perkins is right to hand the palm to the reformers. Their relative victory was apparent in the new breed of almanacs such as *Whittaker's*, advancing a concept of time - and this is central - that was algorhythmic, quantitative and clock-based. Banished to the social and intellectual margins - where it still survives - was the old communal, qualitative time incorporating planetary and lunar cycles, and their corollaries in the annual seasons.

This raises the question of whether a 'post-modern' suspicion of science, ecological crisis in our relations with nature, and a post-Newtonian quantum physics signal the imminence of a new popular sense of time, one that may have significant continuities with premodern cycles and qualities. Whatever the outcome, future historians will have to consult Perkins before setting out.

Patrick Curry

Robin Heath is an astronomer and the author of *A Key to Stonehenge*, Bluestone Press, 1993. He was formerly a senior lecturer in mathematics and engineering and is the founder of Megalithic Tours, Cwm Degwel, St. Dogmael's, Cardigan SA43 3JF, UK.

Norriss S. Hetherington is director of the Institute for the History of Astronomy at the University of California, Berkeley, and the editor of the *Encyclopedia of Cosmology* (Garland Publishing, 1993).

Alan S. Weber is CEMERS Associate Fellow, State University of New York, Binghamton, teaching in the English department. He is currently a Visiting Assistant Professor at Pennsylvania State University.

Ken Negus was for many years Professor of German at Princeton. He is the author of *Kepler's Astrology: Excerpts*, Princeton, 1987.

Martin Bauer is a lecturer in Social Psychology and Research Methodology at the London School of Economics. His research focuses on 'resistance to change' and on the 'public understanding of science'. His publications include *Resistance to New Technology* (ed.), (Cambridge University Press, 2nd ed. 1997).

John Durant is Assistant Director of the Science Museum and Professor of the Public Understanding of Science at Imperial College, London. He is the founder editor of the quarterly journal *Public Understanding of Science*. He is also Chairman of the European Federation of Biotechnology Task Group on Public Perceptions of Biotechnology and is a member of the UK Government's Advisory Committee on Genetic testing.

Patrick Curry is the author of *A Confusion of Prophets: Victorian and Edwardian Astrology* (Collins and Brown, 1992) and *Prophecy and Power: Astrology in Early Modern England* (Polity Press, 1989). He is the editor of *Astrology, Science and Society* (Boydell Press, 1987).

Journal for the History of Astronomy, Science History Publications Ltd., 16 Rutherford Road, Cambridge, CB22 2HH, England.

The Center for Archaeoastronomy, PO Box 'X', College Park, MD 20741-3022, USA. tel: (301) 864-6637, FAX (301) 699-5337. The Center's newsletter also carries news of the International Society for Archaeoastronomy and Astronomy in Culture. <http://www.wam.umd.edu/~tlaloc/archastro/>

Traditional Cosmology Society, Dr. Emily Lyle, School of Scottish Studies, 27 George Square, Edinburgh, EH8 9LD, UK.

British Astronomical Association, Historical Section, Anthony Kinder, 16 Atkinson House, Catesby Street, London SE17 1QU.

Ascella Books, 3 Avondale Bungalows, Sherwood Hall Road, Mansfield, Nottinghamshire, NG18 2QJ, England (reprints of old astrological texts).

Pratum Book Company, PO Box 985, Healdsburg, California 95488, USA, tel (707) 431-2634, Fax (707) 431-0575, E Mail knowledge@pratum.com (extensive range of rare and out of print books on mystical cosmology).

Events:

Astrological Lodge History Seminar, 1 November 1997, 10.00 am-5.30 pm, 50 Gloucester Place, London W1, including Nick Campion on Rudolf Hess' use of astrology, Silke Ackermann on 'Astrology and Scientific Instruments', Judith Kolbas on 'Solar Supremacy or Royal Iconography', Annabella Kitson on 'Lilly's "Mock Suns" and "World's Catastrophe"', Caroline Gerard on the Lauriston Horoscope, Tomas Gazis on Byzantine Astrology. Tickets £10 members, £12 non-members, from Astrological Lodge of London, 50 Gloucester Place, London W1 3HJ.

Culture and Cosmos

CULTURE AND COSMOS

Culture and Cosmos is published twice a year, in spring/summer and autumn/winter.

Contributions and editorial correspondence should be addressed to Nicholas Campion, The Editor of *Culture and Cosmos*, PO Box 1071, Bristol BS99 1HE, UK, E Mail <culture@caol.demon.co.uk>.

Deputy Editor: Patrick Curry, Ph.D.

Editorial Board:
Dr. Silke Ackermann, Professor Anthony F. Aveni, Dr. Guiseppe Bezza, Professor. J. Bruce Brackenbridge, Dr. David Brown, Dr. Charles Burnett, Dr. Hilary M. Carey, Dr. John Carlson, Professor Robert Ellwood, Dr. Germana Ernst, Dr. Ann Geneva, Dr. Jacques Halbronn, Robert Hand, Professor Norris Hetherington, Professor Michael Hunter, Professor Ronald Hutton, Annabella Kitson MA, Dr. Nick Kollerstrom, Dr. Edwin C. Krupp, Dr. J. Lee Lehman, Professor Kenneth Negus, Professor John North, Professor P. M. Rattansi, Professor Francesca Rochberg, Professor James Santucci, Robert Schmidt, Professor Richard Tarnas, Dr. David Ulansey, Robin Waterfield, Dr. Charles Webster, Dr. Graziella Federici Vescovini, Dr. Paula Zambelli, Robert Zoller.

Copy Editor: Ian Tonothy
Technical Assistance: Sean Lovatt

Subscriptions:
Individuals £13 UK and Europe, £15 overseas*
Institutions £22 UK and Europe, £24 overseas
Payment should be by sterling cheque or money order, or Eurocheque, visa or mastercard. For credit card payments we need the full card number, address and name on card, and expiry date.
Subscriptions can be received by E Mail on <subs@caol.demon.co.uk>

*Members of the British Astronomical Association, The Astrological Association and The Historical Association are entitled to a discount. Please enquire.

Contributors Guidelines: Please see inside back cover.
Copyright of signed articles and correspondence remains with the authors.
Copying: Apart from fair dealing for the purposes of research or private study, or criticism or review, as permitted under the Copyright, Designs and Patents Act 1988, no part of this publication may be reproduced, stored or transmitted in any form or by means without the prior permission of the Publisher.

The cover shows the explorer Gonsales travelling to the Moon, from Francis Godwin's *The Man in the Moone* (1618).

Published by Culture and Cosmos, PO Box 1071, Bristol BS99 1HE, UK.
© Culture and Cosmos 1997
Printed by Cromwell Press Ltd., Broughton Gifford, Melksham, Wiltshire SN12 8PH

www.ingramcontent.com/pod-product-compliance
Lightning Source LLC
Chambersburg PA
CBHW071459080526
44587CB00014B/2156